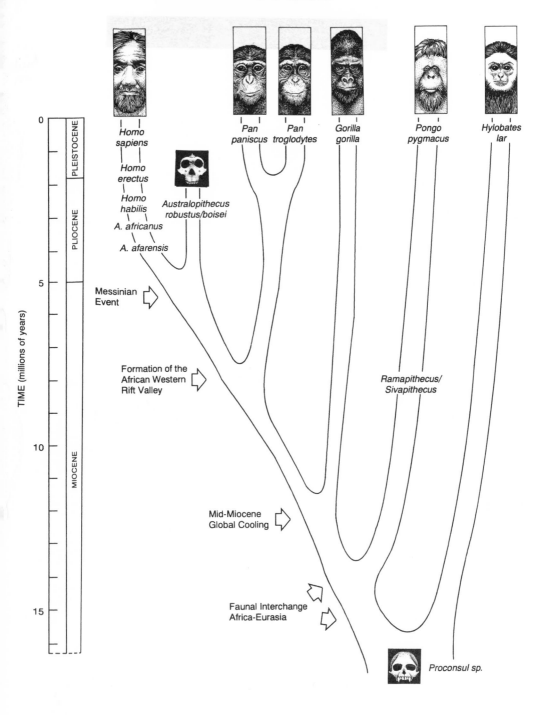

# QUARRY

## CLOSING IN ON THE MISSING LINK

## NOEL T. BOAZ

THE FREE PRESS
*A Division of Macmillan, Inc.*
NEW YORK

Maxwell Macmillan Canada
TORONTO

Maxwell Macmillan International
NEW YORK   OXFORD   SINGAPORE   SYDNEY

The Free Press
A Division of Macmillan, Inc.
866 Third Avenue, New York, N.Y. 10022

Maxwell Macmillan Canada, Inc.
1200 Eglinton Avenue East
Suite 200
Don Mills, Ontario  M3C  3N1

Macmillan, Inc. is part of the Maxwell Communication Group of Companies.

Printed in the United States of America

printing number
1   2   3   4   5   6   7   8   9   10

**Library of Congress Cataloging-in-Publication Data**
Boaz, Noel Thomas.
   Quarry : closing in on the missing link / Noel T. Boaz
      p.   cm.
      ISBN 0–02–904501–0
      1. Man—Origin. 2. Human evolution—Philosophy. 3. Physical
anthropology—History—20th century. I. Title.
   GN281.B62    1993
   573—dc20                                                    93–21689
                                                                    CIP

TO
MELEISA

# CONTENTS

# 1

# Ape into Human

HUMAN EVOLUTION AS UNDERSTOOD
IN THE LATE 1960s AND EARLY 1970s

Anthropology is one of those sciences like astronomy. The phenomena—people, stars—are around us every day. This fact can make for a certain complacency until one stops to ponder the immensity and scope of the questions that surround human and cosmic origins. What could be more commonplace than people and stars, which we see every day and night? And yet what could be more difficult to understand? Of course people and stars have been known for a long time, and no explanations are necessary if one accepts the standard answers that have satisfied generations. All cultures around the world have origin myths to explain how people and stars came to be. They are just there—"that's the way it is," or "God did it, the Bible says it, and I believe it"—end of discussion.

Astronomers and anthropologists have had to overcome some long-held and fervent beliefs in freeing themselves to investigate scientifically the nature of the universe and of humanity within it. Progress was made by key individuals who questioned things. My grandmother Boaz, a God-fearing

1

woman who believed in leaving well enough alone, called such people "doubting Thomases." She loved me dearly, but she realized early on that I was one of them.

Doubting Thomases first arose in astronomy. Copernicus, whose heliostatic ideas replaced the geocentric notion of the universe (and humanity); Galileo, who was forced to recant his heliocentric theory and was put under house arrest for eight years; and Tycho Brahe, who ran afoul of the church when his discovery of a "nova," or new star, shook confidence in the immutability of the universe. These pioneers offered a different interpretation of what ethnologists call *existential postulates*, fundamental ideas in human society about who we are and where we fit into the universe. The heresy that these early astronomers were guilty of was that earth, our home, was not the unchangeable point around which the heavens revolved: We were not in fact the center of the universe.

The key word here is, of course, "we." Astronomers eventually were able to work out a compromise with the church that left out the human or anthropological component. Human beings might not be the center of the universe, but they certainly were the center of their corner of it.

A century or two later Darwin's theory of evolution by natural selection questioned even this most sacrosanct of the tenets of the traditional Western religio-philosophical view of things. People had originated out of nature just the same as every other living species. Darwin thought that "there is grandeur in this view of life," but most others were appalled. Darwin's scientific reasoning and supporting data, however, were compelling, and the theory withstood a firestorm of controversy. Finally, human beings could be studied scientifically—our origins, our history, everything.

I always have been fascinated by the strange paradox that leaves us with a great deal of in-depth knowledge of things of distant relevance but with an imperfect idea of extremely important things close at hand, some of which we think we already know. For example, in school I was puzzled as to why our biology class should be studying the frog in such excruciating detail yet ignoring human anatomy and biology. As I became older I thought of this as a sort of displacement behavior. When direct topics of investigation are too threatening, most people, scientists included, turn to more palatable spheres of activity. Great Soviet minds of the Cold War era turned to chess, for example, when their intellectual pursuits did not fit into the state's plan for applied military-industrial research. Frogs are superficially less relevant and threatening to an anthropocentric view of the world than talking directly about human evolution and human biology. Perhaps the strength of this world view explains why human evolution is such a young science. A science that asks "why" of the most obvious of all phenomena, ourselves, is bound to run afoul of those in society who have serious investments in the status quo. Darwin asked why, and he was pilloried by the public for it.

In this book, I will tell the story of the most recent answers to Darwin's "why?" questions. Why are there human races? Why do individuals differ from one another within one race? When and where did the human species originate? Why did people begin walking on two legs? Why do human beings have such large brains? What role did the use of tools play in human evolution? And on and on.

At first a lot of the research was descriptive. Anthropologists worked on chronicling and classifying the different varieties of humankind. This interest started in the eighteenth century. The German professor of anatomy and physiology, Johann Friedrich Blumenbach, was the first to concentrate on classifying human races on the basis of their skull shapes and physical features. He considered that there were five major races. Blumenbach is considered to be the father of anthropology.

The advent of Darwin's theory allowed the observations of human variability to be organized into a framework. There was a history to the varieties of people that we see today. They did not just spring from the head of Zeus. They had evolved, adapting over the generations to environments as the most fit in the population survived and reproduced and the less fit died or produced fewer offspring. (The rephrasing of Darwin's message by sociologist Herbert Spencer, "survival of the fittest," caught on in the public mind, but it was inaccurate. Survival was not so much the issue as the relative number of offspring that survived and reproduced.)

The truly terrifying idea that Darwin's new theoretical formulation unleashed was that humankind had sprung from the loins of an ape eons ago. "Let us hope that it is not true," one Victorian lady is supposed to have uttered, "but if it is, let us hope that it does not become generally known." Implicit in the idea was that there had been a "missing link," a term coined by the English anatomist Thomas Henry Huxley to refer to that half animal, half human being that some future scientist would find in the fossil deposits of Africa. Huxley's researches were based on detailed comparative anatomical study of the living issue of that early evolutionary split—humans and the great apes. He had determined from meticulous dissections that the African apes, the chimpanzee and gorilla, not the Asian orangutan, were closest to humankind. Darwin agreed. The conclusion followed that "it is somewhat more probable that man's progenitors arose on the African continent than elsewhere," as Darwin wrote in his *Descent of Man* (1872).

The concept of a missing link spread into the popular press. It jumped linguistic barriers. Even the famed German evolutionist Ernst Haeckel threw out the German "vermisste Verbindung" and used the term missing link instead. The original Huxleyan concept related to a fossil form yet to be discovered of a creature that had long been extinct. In the popular mind, there was even a time when the missing link was so misunderstood that people expected it to be found alive in some remote corner of the world.

Many scientists went in search of the missing link, with the idea that simply finding and describing fossils would suffice to answer Darwin's questions. Some even claimed to have found it. Eugene Dubois, a Dutch anatomist and medical officer, went to the Dutch East Indies (now Indonesia) near the habitat of the orangutan to look for fossils that could document the missing link. In 1891 he reported the discovery of a thigh bone and a skull cap of a creature that he named *Pithecanthropus*. *Pithecanthropus*, meaning "ape-human" in Greek, had been Haeckel's name for his hypothetical missing link, which he considered to have inhabited Asia, not Africa. Haeckel's anatomical researches had indicated a greater similarity between humans and orangutans, contra Huxley and Darwin. But a century of research has shown that *Pithecanthropus*, although a more primitive species, is still a member of the human genus *Homo*. *Pithecanthropus* is now known as *Homo erectus*.

Africa did not become the site of serious anthropological investigation into human origins until 1924, and then it was by chance. Raymond Dart, a young Australian anatomist trained in England, had just entered a job as anatomy professor in Johannesburg, South Africa, when a fossil skull of an ape-like creature turned up. Dart, who was more interested in brain anatomy, dutifully if excitedly described the specimen, naming it *Australopithecus africanus*, "southern ape from Africa." But *Australopithecus*, even if it was more primitive than any other human-like fossils then known, was not the missing link either. It already had the anatomy of a lineal ancestor of humans, and it was a bipedal *hominid*, a member of the human zoological family.

In the sixty-five years since the discovery of *Australopithecus* an immense amount of effort has gone into exploration for the true missing link, that fossil creature representative of a population of half ape, half hominids that gave rise to both living humans and living apes.

To know where to look and what to look for, anthropologists have created theories about what kind of animal our common ape-hominid ancestor was: what sort of environments it lived in, how it moved about, and what kind of behavior it exhibited. Some of these theories have been clever and remarkably prescient, utilizing many small clues to derive an overall conclusion that in time has proved correct. Huxley's theory of the African origin of the hominid line, based on details of comparative anatomy of living human beings and living apes, is an example. Many other theories have ranged from novel rearrangements of facts to come up with different conclusions to concoctions of almost pure guesswork, interesting expositions of the possible, but nothing else. The last century and a half of human evolutionary literature is a fascinating excursion into science and science fiction. Our task, however, avoids the first one hundred years of origin theories when the only research technique was description. Instead we concentrate on the most recent forty years. It is not just theories that have changed the modern period, but also the research techniques on which they rely. The interplay between technique and theory is a fascinating, if little understood, phenomenon. It is my hope that, reading this book, you will learn not just what we think we know, but perhaps, more importantly, how we know it. For, truly, science has revolutionized the search for the missing link.

## THE BIRTH OF BIOLOGICAL ANTHROPOLOGY

Modern biological anthropology can be dated from a unique meeting of scientists held in 1950 at Cold Spring Harbor Biological Laboratory on Long Island, New York. Two traditions of analyzing human evolution, variation, and adaptation collided at this meeting with the force of speeding trains, and out of the wreckage crawled modern biological anthropology. I term the two contending factions the "Old Guard" and the "Young Turks." With the hindsight of history, the lines between these two perspectives is razor-sharp. Yet at the meeting there was only a vague comprehension that two different approaches—anthropology and genetics—were vying.

Representative of the Old Guard was Earnest A. Hooton, the erudite, witty, and always nattily dressed professor of anthropology at Harvard who had entered the study of human evolution through the classics. He had been a Rhodes Scholar at Oxford University and had there become enamored with the subject of human origins. Hooton was an inspiring teacher, and he trained most of the pre-World War II generation of biological anthropologists. Among his students, Harry Shapiro went to the American Museum of Natural History, Sherwood Washburn went to the University of Chicago and then to the University of California at Berkeley, Joe Birdsell went to UCLA, Wilton Krogman went to the University of Pennsylvania, and Bill Howells went to the University of

Wisconsin and eventually came back to Harvard to take over when Hooton retired. All of them spoke of Hooton with fondness, as a man who held weekly teas for the students in his home and who took an active interest in their careers. Few could match Hooton's wit, turn of phrase, and output. He dedicated one of his most popular books, *Up from the Ape*, to his critics, "humbly offered for their disapprobation." As his research waned Hooton turned increasingly to the public for approbation. He produced engagingly written books entitled *Young Man You Are Normal*, on normal human variation, and *Man's Poor Relations*, on the nonhuman primates.

Hooton's research Armageddon was Pecos Pueblo, now almost a forgotten footnote in the annals of biological anthropology, buried by a generation of adulating students willing to turn a blind eye to their mentor's failings. Even the usual to-the-point Sherry Washburn pulls his punches on Hooton and Pecos Pueblo.

Pecos Pueblo was an archaeological site in New Mexico excavated by Harvard archaeologist A. V. Kidder. Many human skeletons were found, all carefully excavated and sent back to the lab for analysis. Kidder asked his colleague Hooton to undertake their study. Hooton meticulously measured and described the skeletons, but what landed him in trouble was his interpretation of the results. Hooton made sense of the many variations that he saw in the bones of the Pecos Indians by classifying the skeletons into a number of discrete types. He then postulated a series of invasions or migrations of Asian peoples into North America to account for each of the types. Critics loudly pointed out that Hooton's types were merely arbitrary cuts in continuous variation and that they did not exist. What were types anyway? They certainly did not have any genetic or statistic verifiability. Hooton's clever ripostes supporting his typology kept his critics at bay, but he retreated more and more into the realm of popular writing. At the Cold Spring Harbor symposium of 1950 he appeared as a dapper man, hands on hips as he slipped the jabs of the Young Turks, but his typological research paradigm was collapsing. Soon no one in science would speak of the "Aryan Type" when referring to western

European populations, and anthropologists would desist from christening fossils as new types with uniquely new genus and species names.

Sherwood Washburn was the leader of the Young Turks. He appeared at Cold Spring Harbor as a jaunty, intense young man with open collar. Washburn wanted to put to rest once and for all the typological approach to human evolution. He was tired of the seemingly endless debates about this or that wrinkle on the tooth crown of a particular fossil human specimen and whether it should carry the same name as another fossil. Much of this debate, Washburn maintained, was pointless because living human populations showed much of the same variation within their current boundaries. He believed that anthropologists should fall into line with other biologists and use defensible zoological principles to talk about species and populations. As reasonable as this proposal sounds today, it was radical at the time. Why, for example, should a specialist on birds, such as Ernst Mayr, or a specialist on fruitflies, such as Theodosius Dobzhansky, be at a conference on human evolution? It didn't make sense to the Old Guard. What they did not realize was that these scientists were completing the circle that Darwin had begun to draw around all evolving life on earth, one that included human beings inside.

Collectively, the numerous papers published in 1951 as a result of the Cold Spring Harbor conference probably represent the most important single work in biological anthropology of this century. This body of work represents what the historian of science Thomas Kuhn has described as a "paradigm shift" in science—that time of confusion and unsettledness when scientists are reassessing the fundamentals of their lives' work and research designs. In fact, most scientists cannot summon the personal strength to abandon the frameworks in which they have spent their professional lives. Kuhn points out that they have to die, taking their outmoded research paradigms to their graves, before a new generation, with less invested in the old ways, can move forward. Such was the case here. The Old Guard did not change its collective mind. Its adherents went down with all flags flying, dapper to the end.

The Old Guard basically believed in Hooton-style *types*. One of the participants at the conference, W. H. Sheldon from Columbia University, presented a paper in which he defended the concepts of *somatotyping*—the assignment of a three-digit code to a person on the basis of their body type. He believed that with this code people's behavioral characteristics, particularly criminal behavior, could be predicted. He had so much confidence in his methods that he proposed that "social control" and "controlled human breeding" could result from his findings. These conclusions take biological anthropology out of the realm of the academic and into the realm of public policy, or perhaps science fiction.

Prior to the Cold Spring Harbor meeting this extension of typological biological anthropology into public policy had already been tried. A friend of my family who emigrated to the United States with her Russian husband after World War II remembers having her head measured by white-coated German scientists as a young woman in occupied Poland. Those with head shapes that conformed to a specific cut-off point determined by the German anthropology professors were classified as Aryan types and were sent one place; those with slightly different head shapes that were below the cut-off point were sent to concentration camps. She survived the cut-off point, but several other millions in Europe before and during World War II were not so fortunate. Typological anthropology served as the theoretical foundation for the most radical of modern social control movements—the extermination of non-Aryans perpetrated by the National Socialist Party that ruled Germany from 1933 to 1945.

The Young Turks at Cold Spring Harbor objected strenuously to the typological approach of Sheldon and Hooton by pointing out that individuals within populations were variable and that Sheldon had no proof that his ideal types were based on any real genetic foundation. Washburn maintained that somatotypes could change from parent to child depending on the environment. He thought that typing of individuals should be "replaced by getting some understanding of the processes which cause the differences." Sheldon disagreed, saying that description should come first, but that this undertaking had seemed "to impose too great a burden on the human mind."

The biggest bombshell dropped on the Old Guard, however, came from Ernst Mayr, a German-trained ornithologist and specialist in the naming (taxonomy) of species in nature. Using the new yardstick of variability within populations, he stated that "after due consideration of the many differences between Modern man, Java man, and the South African ape-man, I did not find any morphological characters that would necessitate separating them into several genera." He suggested that *all* the fossil human-like specimens that anthropologists had discovered after so much laborious effort over the preceding century be simply ascribed to *one* genus, our own—*Homo*. In other words, the entire "Age of Description," from before Darwin to Cold Spring Harbor, was a waste of time. His opinion was that the differences were not as great as between genera of other animals. This assertion meant that the wonderfully diverse lexicon of human paleontology, a virtual linguistic playground for the classically educated, with melliferous names such as *Plesianthropus transvaalensis*, *Meganthropus palaeojavanicus*, *Africanthropus njarensis*, *Sinanthropus pekinensis*, *Pithecanthropus erectus*, and so on, were to be replaced. Everything was now to be simply *Homo*, with three species: *Homo transvaalensis*, *Homo erectus*, and *Homo sapiens*.

Mayr's proposal went so far that even Washburn argued that at least the South African *Australopithecus* be retained (instead of *Homo transvaalensis*) because it showed such significantly more primitive anatomy than members of the genus *Homo*. Mayr simply countered that the population is the important unit of evolution and that the population is what the species designates. How one determines a genus is arbitrary. The definition is gauged by the relative amount of difference that one sees between the genera of other animals and, in Mayr's opinion, hominid fossils don't show very much difference. To anthropologists, this statement was a bit like telling a new mother that her baby looks like every other baby. It did not go over well.

To this day most anthropologists have not accepted Mayr's

suggestion. But Mayr's paper and the debate that ensued at Cold Spring Harbor in 1950 established the pattern that has characterized anthropological debates over naming of fossils ever since. A sort of two-party system has developed. On one side of the aisle sit the "lumpers," who, like Mayr, prefer to emphasize the similarities in fossils and to lump them into large categories. They debate the "splitters" across the aisle, who focus on the detailed differences between each and every fossil and tend to give names based on those differences. After Cold Spring Harbor, lumpers and splitters did not like each other any better, but they at least had a common proving ground—they had to show that the species that they were proposing for fossils compared closely with the variability seen today in living species. Although this paradigm represents a major step forward from the idealized types of earlier anthropologists, it has not stemmed the tide of debate over hominid taxonomy, as we shall see.

Prior to Cold Spring Harbor, scientists who study the human body and its evolution were known exclusively as "physical anthropologists." Shortly thereafter, Washburn proposed the names "experimental physical anthropology" and the "new physical anthropology" to describe the now-transformed discipline, but the term "biological anthropology" increasingly has come to be used. It emphasizes how much the field of human evolution has now become a synthesis of the traditional subject matter of anthropology and the theory of biology. As this book progresses, we will take up particular questions of biological anthropology, such as: What can we learn from the history of behavior? What can we learn from changing climates? What do tools tell us about changes in mental capacity?

OFFSPRING OF THE NEW FIELD

The major fields of study that biological anthropologists engage in took shape in the 1950s in the wake of the Cold Spring Harbor conference. The term *paleoanthropology*, first proposed by the French anthropologist Paul Topinard in the late 1800s, was resuscitated by M. F. Ashley Montagu at the

Cold Spring Harbor meeting. Paleoanthropology connotes a much broader field than "human paleontology," which is the study of fossil human bones and the indications they give for evolutionary lineages. What is now considered the old core of paleoanthropology, the naming of fossil species and the interpretation of lineages, still evokes much of the heat in debates, but as the field has matured systematic investigations into early hominid environmental and chronological contexts, functional anatomy, and behavior have quietly become the norm.

Paleoanthropologists now aspire to understand the *function* of the anatomy of fossil bones, to bring them back to life, to place them into the behavioral repertoire of the ancient hominid to which they belonged. They are concerned with context—exactly how old the bones are, what sort of environment the hominids lived in, what animals and plants shared that environment, whether they ate or were eaten by the hominids, and what the associated archaeological remains such as stone tools or structures may tell about ancient hominid behavior.

This approach to paleoanthropology was launched by another Hooton student who was at the Cold Spring Harbor meeting, Joe Birdsell, then still a graduate student. In 1953 Birdsell teamed up with a biologist, George Bartolomew at UCLA, to produce a paper entitled "Ecology and the Protohominids." It investigated the environment, animal and plant interactions, and behavioral contexts of hominid evolution from the standpoints of modern biology and ecology. This paper established a new subdiscipline known as "paleoecology" and spawned a spate of new hypotheses about early hominid divergence.

Suddenly anthropologists became aware that there was a new playground for ideas, and almost everyone got into the act. Linguists hypothesized on the origin of language; some thought it started with gestures, some thought that the australopithecines sang before they talked, and some tried to tie toolmaking to speech. Other anthropologists came up with the idea of a "cultural ecological niche" for hominids and, echoing Ernst Mayr's paper at Cold Spring Harbor, they suggested that no more than one hominid species could have existed at any one time in the past. This idea developed into the "single species hypothesis" of Loring Brace and Milford Wolpoff at the University of Michigan. There was a plethora of new ideas on the evolution of upright walking or bipedalism: early hominids needed to see over tall grass on the African savanna; they stood up in order to carry food back to their campsites; higher stature gave them a more dominant ecological position in the food chain compared to other predators; by standing up their hands were freed for tool use, and so on. There was a flood of such hypothetical papers throughout the 1960s and early 1970s. This research had one common thread—the hypotheses were framed in biological and ecological terms. The missing link was now thought of as a population of animals, not one individual.

Paleoanthropologists, who now had a new paradigm, soon found that they had too many hypotheses and not enough data. More attention and more grants began being directed to research into the earliest phases of hominid evolution in Africa. There had to be some weeding out of the hypotheses. Data were needed to test all the widely divergent ideas of how hominids had come to be.

The efforts of one lone paleoanthropologist out on the African savanna then began to enter the mainstream. Louis Leakey was not impelled into the African heartland by theoretical formulations. He had been born there. He had struggled since 1931, largely unfunded and largely unsuccessfully, to discover fossil evidence of early hominids in Africa. In 1959 he and his wife Mary finally discovered the dramatically complete hominid skull for which they had been searching for twenty-eight years. Their work at Olduvai Gorge, Tanzania, was used in elementary science classes the world over as an example of the need for persistence in scientific research.

Still untouched out in East Africa by the influences emanating from Cold Spring Harbor, Leakey named his 1959 skull a new genus and species, *Zinjanthropus boisei*. The genus name was in honor of the old Arabic term for East Africa, "Zinj," plus the Greek "anthropus" for "human or man," and the species name was in honor of one of Leakey's benefactors in England. Most other anthropologists then considered and still now recognize the skull as a member of the hominid genus *Australopithecus*. But the taxonomic squabbling that surrounded Leakey's discovery was drowned out in the chorus of accolades. Here was a natural laboratory to test all the ideas that were pouring out of academia.

One young disciple of Washburn's who had also been at Cold Spring Harbor, F. Clark Howell, then of the University of Chicago, lost no time in getting to East Africa. He arrived in Nairobi just months after the *Zinjanthropus* discovery. After consultations with Leakey he headed north in an exploratory reconnaissance of the Lower Omo Basin of southern Ethiopia. Fossils had first been discovered at Omo by an Austro-Hungarian expedition led by Count Samuel Teleki in 1888, and French paleontologist Camille Arambourg had made significant fossil collections in 1932–1933. Leakey himself had sent a Kenyan collector to Omo during World War II. But despite the discovery of many excellent fossil specimens no hominids had been found.

Howell's trip had been hastily planned and mounted. He travelled alone in a Landrover that he discovered en route had a cracked engine block. Finding abundant fossil-laden deposits, he made some representative collections, only to have them confiscated by Ethiopian border authorities as he was heading back to Kenya. Years later fossil specimens of unknown origin were noticed paving the driveway to the district governor's house in the southern Ethiopian outpost of Kalam; they were probably Howell's. Yet Howell had succeeded in ascertaining that there was potential at Omo and he went back to Nairobi to report to Leakey.

Meanwhile Leakey's own site, Olduvai Gorge, located in northern Tanzania, began to yield its potential, due in no small part to increased levels of funding. As momentous as the discovery of *Zinjanthropus* had been for Leakey's scientific career, equally momentous was his discovery of American funding sources. The National Geographic Society was primary among them. With reliable vehicles and more funds for workers' salaries, excavations and field surveys at Olduvai were extended. In 1968 Leakey founded with some backers in southern California the L.S.B. Leakey Foundation, based at Pasadena.

During the early 1960s a series of discoveries at Olduvai resulted in the recognition of a new species of hominid, *Homo habilis*. The new species name was coined in 1964 by Louis Leakey and two biological anthropological colleagues, Phillip Tobias of South Africa and John Napier of London. *Homo habilis* was exactly what Leakey had been looking for for decades: the very first demonstrably human ancestor, associated with stone tools and showing the human-like anatomical features of increased brain size, reduced overall dentition, and nonprotruding face. Earlier he had been all too ready to plug *Zinjanthropus* into this role, despite its very unhuman, massive chewing apparatus and the relatively small brain size of what was nicknamed Nutcracker Man. Previously, in the 1930s, Leakey had made similar claims based on a fossil jaw found under unclear geological conditions at the western Kenyan site of Kanam.

*Homo habilis* was the first early hominid species to be named in accordance with the biological tenets proposed at the Cold Spring Harbor meeting in 1950. But perhaps because of the earlier claims by Leakey, *Homo habilis* became a point of significant debate in the paleoanthropological community. Splitters, such as Howell, accepted the name as distinguishing a different species of early hominid. Lumpers, such as Brace and Wolpoff, considered the new fossils from Olduvai a part of either the more primitive *Australopithecus africanus* or the more advanced *Homo erectus*. When I first met Louis Leakey, seven years after *Homo habilis* had been baptized, the question that I posed to him was how did he respond to critics who questioned the distinctiveness of the species and on what evidence did he base his rebuttal. As a student immersed since prep school in the cut and thrust of academic debate I was unimpressed with Leakey's dismissive answer. "The variability that has led some workers to question the validity of *Homo habilis*," he sniffed, "is not a view to which I subscribe." As a bejewelled lady of the Washington social set led Leakey off to his next engagement, I stood there pondering the fact that I had just met one of the Old Guard, one of the true typologists.

A year later, in 1972, Leakey died of a heart attack in London on his way to another lecture tour of the United

States. Now, looking back with the perspective of twenty years of hindsight, there is reason to view Louis Leakey more as an important transitional figure than as exclusively a member of the Old Guard. If Sherwood Washburn transformed the theoretical underpinnings of modern biological anthropology, Louis Leakey transformed the empirical methods of several of the important parts of the discipline. Washburn had the new ideas and Leakey had the new data to test them.

## GEO-SCIENCE ENTERS

One of the most important developments in paleoanthropology that Leakey had a hand in was the integration of new methods of dating fossil sites. Accurate dating of geological strata, fossil bones, and archaeological artifacts was a natural outgrowth of the new experimental physical anthropology. With independently measured dates one could cut through the circularity of argument that assessed age of a deposit on the basis of the evolutionary stages of the animals in it and then used the age to investigate the evolutionary history of the animals.

The era of radiometric dating was ushered in with Willard Libby's development of the carbon-14 dating process, for which he won the Nobel Prize in 1955. The method was based on the atomic theory. Carbon-14 was formed in the upper atmosphere by solar radiation and was two neutrons heavier than carbon-12, the more common variety. It was not stable, however, and calculations showed that it would slowly degrade, or decay, to carbon-12. What was exciting to physicists was that this rate of decay was constant. When any animal or plant died and no more carbon-14 was entering its system from the air, half of the carbon-14 would be gone in 5,500 years. The amount would be halved every 5,500 years until there would be too little to measure. This limitation was usually about 50,000 years, and only recently have technical advances in measurement allowed fossils older than this to be dated by the Carbon-14 method. Paleoanthropologists were quick to grasp the potential of the new method, and soon after Libby had set up his new laboratory in Chicago he was running dates on archaeological sites, funded by an anthropological foundation, the Wenner-Gren Foundation in New York.

Leakey's fossils were much older than 50,000 years, but no one knew exactly how old. Two geochemists at the University of California, Berkeley, Jack Evernden and Garniss Curtis, proposed to use a different type of dating method, one that had a much slower decay rate than carbon-14. This technique, the potassium-argon method, uses the rate of potassium decay to argon gas as the clock to measure the age of rocks. The half-life of potassium is several million years, so it is well suited to dating older rocks. Volcanic rocks have a lot of potassium and they occur in abundance in eastern Africa, as lavas, basalts, or ashes (called "tuffs"). Evernden and Curtis took samples of these rocks from Olduvai Gorge and subjected them to potassium-argon analysis.

What Evernden and Curtis found astounded even Leakey. Bed I at Olduvai, the level from which both *Homo habilis* and *Zinjanthropus* had come, was well in excess of the one million years originally estimated for it. It was almost twice as old, dating to 1.8 million years B.P. (before present).

A masterful promoter, Leakey lost no time in weaving this new-found, high-tech antiquity into his pitch. He had found the earliest human and Africa was now confirmed as the "cradle of humanity." On a more scientific plane, Leakey's application of potassium-argon dating to paleoanthropology ushered in a new age. There was no longer protracted debate about this or that stage of evolution as read from the fossils found associated with the hominids. It was now possible to speak in absolute terms about the age of fossil finds.

The use of the new method spread like wildfire to other fossil sites in eastern Africa, all the ones with volcanic rocks. Unfortunately, there were no dateable potassium-rich rocks in the South African cave sites, where the original *Australopithecus* discoveries were made. Their absolute dates are still matters of debate today. The old method of faunal comparison—which fossil antelope or pig species matches the fossil from your site, and how the evolutionary lineages of those species stack up to give a trend from older to younger—is how the South African sites have been dated.

Dates are important because paleoanthropology is a historical science. What happened when and who the actors on stage
were at any particular time in the past are critical pieces of
information for reconstructing human evolution. Absolute
dates, those with numbers of years attached to them rather
than words such as "early Pleistocene," allow hominid fossils to
be placed precisely on the giant tally board that paleoanthropologists keep. The age of a specific fossil allows a scientist to
place it within a matrix of time and anatomical similarity to the
other hominid fossils known. A fossil skull that is anatomically
similar to others but is much earlier in time, for example, can
be interpreted as representative of a likely ancestral population. A fossil skull that is very dissimilar from others dating to
the same time, on the other hand, is likely representative of a
separate lineage.

The development of accurate dating methods was also necessary for the new focus on ecology that had developed as an outgrowth of Washburn's new physical anthropology. One could
now compare sites, floras, faunas, and hominid species from
one area to the next within a clear time framework.
Paleoanthropologists now began to gather the data to answer
questions such as whether early hominids lived in the savanna,
when they may have left the forests behind in their evolution,
what types of food they had available to eat, and so forth. The
Leakeys' site at Olduvai was the first early hominid site dated
with the new absolute dating method of potassium-argon. It
became the first controlled laboratory for testing many of the
hypotheses that had been developed following Cold Spring
Harbor.

## EVOLUTION OF BEHAVIOR

Anthropologists now are interested in understanding not just
how the fossils of our ancestors and relatives fit into evolutionary lineages but also how the animals from which the fossils
came behaved: what they did, how they moved, what sort of
groups they lived in, and so forth. But much of the past behavior of early hominids that anthropologists thirst to know about

would not have been preserved as fossils. For example, there is no fossil record of sitting and gazing at a beautiful sunset or for what vocalizations australopithecines made when they woke up in the morning. Unfortunately, no matter how spectacular the discoveries at Olduvai or other fossil sites, they could never give us this information. "Behavior does not fossilize," Washburn said. Paleoanthropologists can only find the results of behavior, such as footprints or stone tools, or they can study the anatomy of the fossil bones themselves and determine the range of movements that an early hominid had. The joints of the body are designed to move in only certain directions and this information can be very revealing if analyzed by an astute specialist. But as valuable as this information is, it does not help in determining social behavior, communication, and other daily behaviors that are crucial to know if we are ever to understand the evolution of human behavior.

Washburn's solution was to study the behavior of humankind's closest living relatives, the higher primates. This research could not be conducted by observing the animals in zoos; it would be necessary to live with them in the real world, where they behave naturally. The underlying theory is that primate behavior, including human behavior, is determined by biological and mental heritage on the one hand and by environment on the other. If Washburn could find a close match among the primates in terms of genetic closeness to hominids and in the most similar environment to early hominids, he reasoned that he would have the best chance of being able to deduce something significant about early savanna-living hominids. He chose the savanna baboon.

In 1957 Washburn and a graduate student, Irven DeVore, embarked on a study of the baboons living almost in Louis Leakey's backyard, Nairobi National Park. Their study revealed that baboons live in harems; a single male is surrounded by several females, juveniles, and infants. There is a clear dominance hierarchy—also termed a pecking order from the early behavior studies of birds. Certain males have first access, before other males, to sitting places and to food. Males are usually dominant over females, although females have their own dominance hierarchy. The males are territorial, defending their space from other males with fierce, open-mouthed threats that show their long, impressive canines. They sit open-legged at the corners of their territories with erect penis to advertise their vigilance. The males band together to defend the troop against threats by predators, such as lions or leopards. Their social defense is effective, but Washburn and DeVore hypothesized that without the large canines early hominids probably would have needed weapons to ward off the predators.

The jump from baboon to early hominid was clear and persuasive. Many of the details of early hominid life on the savanna not supplied by the fossil record were filled in by recourse to the behavior of living baboons in that same savanna environment. The principle that study of nonhuman primates in the wild could contribute substantially to understanding human evolution was established.

The point was not lost on Louis Leakey, who, shortly after Washburn and De Vore had finished their second study of Kenyan baboons in 1962, enlisted Jane Goodall to begin her landmark study of chimpanzees at Gombe, Tanzania. Leakey, who prior to contact with Washburn had had no interest in primatology, became one of its great proselytizers. He subsequently promoted the major studies on the remaining two great apes: Dian Fossey's study of mountain gorillas in Rwanda and Biruté Galdikas's study of orangutans in Indonesia. These studies have provided the first long-term data that anthropologists have had on ape behavior in the wild. Goodall's study is the longest continuous behavioral study of any animal species under natural conditions. This work has had a substantial impact on how we interpret early hominid behavior and ecology.

The interaction between Sherwood Washburn, the small, precise New England Yankee professor, and Louis Leakey, the robust, sun-weathered British ex-colonial bush explorer, was incongruous yet critically important in understanding the development of modern biological anthropology. Leakey took Washburn's ideas on experimental anthropology, rolled up his sleeves, and dived in. In one of the early *National Geographic* articles on Olduvai, Leakey is shown immersed in the carcass of an antelope, which he is skinning as an early hominid would have, with a stone tool. In another case Louis Leakey and his son Richard defended a fresh kill from a pack of hungry hyenas on the Serengeti Plain using only sticks, to see if it could have been done by early hominids. Leakey believed that hominids were (and are) naturally distasteful to large cats and thus would have been protected from their onslaughts, but hyenas, in his opinion, were a real threat. In this case, as indeed for millions of years prior to this experiment, it was a dead heat. Some of the meat went to the hyenas, and some was kept by the hominids. The Leakeys escaped unharmed.

Having read this far, it may seem as if for every anthropologist there has been a theory of evolution. Huge fights have arisen over names, places, and dates. Every other paragraph of the last several pages introduced a new faction.

Anyone would be confused. Paradigm shifts—in this case from descriptive to biological anthropology—are confusing. In the past twenty years, however, much has been sorted out. And I have been lucky to see some of this sorting, which is what I hope to share with you in the remainder of the book.

---

I walked into Sherwood Washburn's office at the University of California at Berkeley in June 1972 when I was twenty years old. The paradigm shift in biological anthropology was only slightly older. Washburn had become the dominant force in American biological anthropology, first at the University of Chicago and after 1960 at Berkeley. By 1972 Washburn had built the premier program in human evolution and that was why I had come. While waiting to see him, I reviewed in my mind the sequence of events that had led me to his office.

I had been obsessed with the idea of human evolution since the age of twelve, when my prior interest in dinosaurs was transferred to prehistoric people by a book that I read in the sixth grade. Two years later, having exhausted all locally available sources, I began to look for schools that could give me courses in anthropology.

Anthropology then, as now, was not a widely taught course

in secondary schools in the United States, but I was able to find two schools, Phillips Exeter Academy in New Hampshire and Verde Valley School in Arizona, that offered good programs. I went to summer school at Exeter in 1967 and took an intensive anthropology course with Zdenek Salzmann, a Czech anthropologist with a doctorate from Indiana University who specialized in Arapaho, an American Indian language. Salzmann, who preferred to be called Denny by the students, introduced me to the use of calipers, one of the standard measuring tools of the trade for biological anthropologists, and casts of fossil hominids. I had learned each of the specimens in the case in the anthropology room at Exeter within the month. But then, as throughout my subsequent study at Exeter, Harvard University Summer School, and the University of Virginia, my early training in anthropology was dominated by cultural anthropology. I learned all about Crow and Omaha kinship systems, Australian aboriginal subincision, uxorilocality among the Benda of Zimbabwe, and the sexual mores of New Caledonians.

Despite my heavy dose of cultural anthropology, I was happy as an undergraduate at the University of Virginia. The university allowed me to put together my own major, a "university major" in physical anthropology. I was able to put into my course of study medical school gross anatomy and neuroanatomy, human genetics and molecular genetics, ethology, and archaeology. Altogether, this curriculum was an excellent training for graduate school, but I never took a course in physical or biological anthropology. My mentor was Dr. Charles Kaut, an anthropologist from the University of Chicago who specialized in Apache and Philippine Tagalog kinship systems. Kaut reminded me of Ulysses S. Grant. He was a bearded, hard-talking, cigar-smoking, hard-drinking former Illinois farm boy.

Charles Kaut became my conduit to Sherwood Washburn. Kaut had been president of the graduate students' Anthropology Club at Chicago, an organization for which Sherry Washburn served as faculty advisor. He had hung around with the physical anthropologist graduate students, too; Washburn's star graduate student, Clark Howell, had been

Kaut's roommate. During the 1950s Chicago was the place to be for all branches of anthropology. Kaut had absorbed the state of the art in biological anthropology just by being around many of the major players in the field, and he relayed this information to me in the form of daily discussions, stories, and suggested readings in his office, which he let me use to study at night.

By the time I had finished my second year of undergraduate school I had, under Kaut's tutelage, read virtually every one of Washburn's papers as well as his edited books. By the time I finally met the man I felt that I had already known him for years. I had a pleasant, nonsubstantive discussion with Washburn, saying that I was at Berkeley for summer school, before my last year at the University of Virginia, and that I planned to apply to graduate school at Berkeley the following year. I was glad to be taking my first course in physical anthropology, a summer school course taught by a visiting primatologist. What struck me about Washburn were the incongruities: he seemed such an unassuming man to have effected such a revolution in the science of human origins, well known for its towering egos; so taciturn in conversation when his printed works were so voluminous; such a New Englander (although he did pronounce his *r*s) in the midst of the golden girls and surfers. When I left Washburn's office I had a strong premonition that I would come back to Berkeley, but I did not have a clear idea of how I was to interact or work with Washburn himself. Despite the huge debt that I owed to him for my theoretical background, it was with Washburn's star paleoanthropological student, F. Clark Howell, under whose aegis I was to return to Berkeley.

Historian of science Thomas Kuhn in his *Structure of Scientific Revolutions* pointed out that after a paradigm shift in a science there ensues a period of methodical data-gathering and sifting using the new paradigm. After the Washburnian paradigm shift in biological anthropology, Clark Howell represents this second phase of the development of the discipline. Although as an undergraduate at Chicago he was interested primarily in history and jazz, he fell under Washburn's spell and had finished his doctorate by 1953 on a topic Washburn had supervised—split-line analysis of the base of the human skull. This technique, which fit into the new experimental paradigm, was aimed at understanding the meaning of the complicated pattern of bumps, grooves, pits, and foramina on the base of the skull. One softened up the bone with a decalcifying acid bath, then carefully separated with a dissecting needle the long strands of bone known as osteons. The pattern of splits between the osteons was called the split-line pattern and could be interpreted in a variety of ways indicative of function or genetic relatedness.

Howell, however, did not warm to a career in experimental anatomy. After finishing his doctoral dissertation he never returned to the subject, never cited it in his published papers, and never to my recollection ever mentioned it in conversation. Howell instead entered anthropology through his interest in history. He said the idea fascinated him when he learned that there was a time "before history"—prehistory. Howell attacked the subject of paleoanthropology with a passion. He had to know everything, to get every reference, to know all the details. He studied specimens in museums, amassing voluminous notes, observations, and measurements. He also began an active field excavation program to obtain more data to fill in where the museum collections left off. Here Howell was to make his greatest methodological contribution to the field: developing the modern, multidisciplinary paleoanthropological research expedition. Finally, there would be a chance of obtaining data necessary to test some of the behavioral hypotheses that had come out of the new physical anthropology of Washburn.

While Clark Howell, along with Louis Leakey, was working primarily on the paleoanthropological side of biological anthropology—trying to understand human evolution and behavior from fossils—others had taken Washburn's lead and gone in the direction of comparative ethology by trying to understand human evolution and behavior from behavioral study of living animals. Assessing the theoretical formulations of early human behavior from this standpoint is important because a comparative approach can help to furnish models for the missing links in hominid behavioral evolution, models that paleoanthropology can then seek to investigate through the fossil record. It is to these hypotheses and speculations that we now turn.

# 2

# Naked Apes
# and Killer Ape-Men

The first and most interesting question of behavior concerns (you guessed it) sex. Although biological anthropologists rediscovered the importance of studying sex in the 1950s, I am relieved to report that some of my predecessor anthropologists had managed to stay interested in the subject, from as far back as Darwin's day.

## SEX AND EVOLUTION

In Paris in the late nineteenth century the whole world was like a candy shop full of bizarre cultures and strange peoples waiting to be discovered. It was a heyday for anthropological data-gathering, with curiosities from around the globe displayed and analyzed at museums, the university, and the Academie Française. One day an academician presented his findings that human beings, alone of all the primates, lacked a penis bone. The *os penis*, even in the closely related great apes, runs the length of the penis and gives it a rigidity that the human penis is incapable of achieving. This clearly demonstrates, so the hypothesis went, that human beings have adapted to civilization and have lost much of the animal lust that characterizes the other primates. This conclusion did not go uncontested. Another anthropologist countered that the human penis is also the largest in all the primate order. Why should this be so if, as the first hypothesis averred, sex is so unimportant in human life?

There was no conclusive ending to the debate. Two apparently contradictory facts of comparative human sexual anatomy had been brought to light. Human males have both the largest and the only boneless penises among the primates.

Female sexual anatomy occupied the academicians a few years later when a woman from the Hottentot tribe of southern Africa arrived in Paris. Adult females of the Hottentots and related populations have a protruding enlargement of the fat deposit on their buttocks, a condition known as *steatopygia*. The Hottentot woman looked remarkably like ancient "Venus" figurines that recently had been unearthed from Pleistocene cave sites in France. Here was a chance, the French academicians thought, of being able to examine a "living fossil" of humanity. The woman was met with intense curiosity but a callous public viewed her as merely a newly discovered natural history specimen. She was exhibited around Paris, and when she died a few years later she was stuffed and put into the Musée de l'Homme. She is still there, back in the collections in an ancient glass case among the bones of Neanderthals.

Why did she have such enlarged buttocks? Some suggested that it was for energy storage. But the explanation that received more support was that the steatopygia had more to do with sexual attraction, a conclusion not difficult for men to arrive at who were surrounded by women wearing bustles and corsets. The interpretation that enlarged buttocks related to a heightened degree of sexual activity in "primitive" peoples was strengthened when it was discovered after her death that the Hottentot woman had enlarged *labia minora*. The labia were stretched in Hottentot women and artificially enlarged at puberty as a sign of attaining womanhood, but at the time the condition was thought to be a normal part of female Hottentot anatomy.

Both nineteenth century Parisians and their modern counterparts also have wondered why the human female breasts are so large. Most of the breast is fat, not related to milk production for suckling infants. The ancient Venus figurines also had large pendulous breasts. Did the form and size of the human breast also have something to do with sexual attraction? Comparative primate anatomy showed that the relatively large body size of human females did not explain the large size of their breasts. Gorilla females have quite small breasts and yet effectively suckle their young. Here also was something unique about human sexual anatomy.

Not until the 1960s did broad anthropological interest return to the topic of the relation between human male and female anatomy and sexual behavior. Books by Bernard Campbell, Ashley Montagu, Desmond Morris, and others attempted to explain in evolutionary terms the rather unusual human pattern of sex-related anatomy and sexual behavior. More recent books by Helen Fisher and Lionel Tiger have continued this effort.

One of the most striking aspects about human females is that they do not have a period of heat. Almost all other female mammals, and most primates, have periods called *estrus* when their behavior, their smell, and sometimes even the form and color of their external body parts signal that they are ready to copulate and reproduce. Human females on the other hand conceal their monthly period of ovulation, when they are most likely to conceive if sexual behavior takes place. Evolutionarily speaking, it is as if human females are saying that they do not want to reproduce. How could evolution have produced such an unusual adaptation? Why would females that hid the time of their ovulation ultimately survive and have more offspring, the essence of evolutionary success, than females that exhibited the more primitive pattern of externally communicated estrus?

To answer this question we must look at the other unusual aspects of human sex. Men and women reach an excited physiological state of hyperventilation when they copulate. This state culminates in a muscular spasm known as orgasm. Human copulation lasts for a much longer time than nonhuman primate copulation. It is apparently much more pleasurable as well.

Orgasm in humans is related to anatomical changes in the orientation of the female vagina. The vagina opens more forward in women than in nonhuman primate females. The more posterior orientation in nonhuman primates is due to the fact that males in almost all other species mount females from the rear. In humans, on the other hand, face-to-face copulation is much more common. This face-to-face mode brings the clitoris, the highly sensitive female anatomical analogue of the male penis, into a position where it is pushed against the pubic bone during copulation. It receives much more stimulation than is the case in a rear-mount. This stimulation is a major contributor to female orgasm.

Sex clearly means much more to humans than merely reproduction. As marriage counselors can attest, it is bound up in a complicated psychophysical network of feelings that are basic to the entire relationship between a man and a woman. Indeed, this is the best hypothesis to explain why the whole system of human sexual behavior evolved in the first place. Natural selection acted to cement the pair-bond between male and female hominid that is basic to the human social unit. Sex became much more than a means of procreation; it became an important element in ensuring the solidarity of the social unit, which in turn fostered the survival of the species.

With this in mind we can explain why ovulation in human females became hidden. Sex became an important emotional bonding mechanism between male and female hominid, not an activity restricted to a period of estrus once a month. A female who copulated with her mate frequently or whenever he came back to camp from a hunting foray built a strong interpersonal bond. She also had a higher chance of conceiving offspring and having the resources that the male provided to care for herself and the offspring. Females with a marked period of estrus, however, would lack the ability to bond this way with a male and thus would have fewer food resources and would be without the protection that a resident male would provide. She would have fewer offspring that survived and reproduced themselves. In this way, natural selection would weed out individuals in the population that showed estrus behavior.

The bonding hypothesis explains the apparent contradiction in male sexual anatomy. A more sensitive penis lacking an internal bony strut makes evolutionary sense for a species that puts an emphasis on slow, pair-bonding sex. A larger penis is clearly of evolutionary advantage from the standpoint of maximizing female and male sensation during copulation.

It is probably no coincidence that the pair bonding hypothesis was developed in the 1960s, when the sexual revolution encouraged explicit discussion of sexual subjects long considered taboo in society. And it is similarly no accident, in those post–Cold Spring Harbor years, that another intellectual trend was developing, that would have a profound effect on hypotheses on the evolution of human behavior. This trend was ethology, the study of behavior of animals in their natural, wild states from an evolutionary standpoint.

## ETHOLOGY, THE NATURALISTIC STUDY OF BEHAVIORAL EVOLUTION

Ethologists such as Konrad Lorenz of Austria sought to understand the elements of behavior in the natural world and how they had evolved. Lorenz had spent most of his research career with birds, specifically the greylag goose. He was the scientist who had made the concept of pecking order a common term. He used the term to refer to the pattern of behavior in bird societies that resulted in certain individuals consistently being pecked and displaced from feeding areas, while other individuals consistently did the pecking. Human beings were quick to see parallels in human society, from corporate ladders to Boy Scout troops. Other ethologists modified the term to the more general "dominance hierarchy."

What made it scientifically defensible to extend observations and generalizations made on animal behavior to humans? Certainly everyone could recognize the wonderful discoveries of ethology: Von Frisch's discovery that bees communicate complicated information on the location of food sources through "dancing"; Tinbergen's discoveries on the complex point and counterpoint of courtship rituals in sea gulls; and Lorenz's elucidation of the phenomenon of "imprinting" in young greylag geese, in which he was followed as their mother.

But what did these animal behaviors have to do with people? Tinbergen extended his discoveries about animals to treating autism in human children, and Lorenz related his research to dealing with the problems of human aggression. But the implications for understanding human behavior were much more profound than these beginning forays. In 1973 the Swedish Academy recognized these implications, and Lorenz, Tinbergen, and Von Frisch received the Nobel Prize for Physiology or Medicine for their work.

The fundamental connection between animal behavior and human behavior is the theory of evolution by natural selection. Modern ethology is defined as the "evolutionary study of behavior." If we accept that humanity is an outgrowth of the natural world and that we are the product of the same evolutionary laws that produced animal species, then understanding the laws of animal behavior naturally can tell us an immense amount about the laws that affect human behavior. How animals that live in social groups interact and behave is particularly relevant to humans.

Ethologists approached the problem of how early hominids had behaved on the African savannas in a different way than did the anthropologists. Anthropologists Sherwood Washburn and Irven DeVore had studied in the savanna social groups of animals closely related to humans—baboons, on the assumption that all primates are close to humans. Some ethologists disagreed and argued that baboons are probably quite different from early hominids. Baboons do not derive a significant proportion of their food from meat, gained by hunting and scavenging. Early hominids did, judging by their stone tools and cut bones.

Ethologists George Schaller and George Lowther undertook a study of the carnivores in Serengeti National Park in

Tanzania and published the results in 1969. They took the novel approach of viewing early hominids less as primates than simply as one of the many social carnivores that lived out on the savanna. They suggested that more could be learned about early hominid behavior from studying animals adapted to a similar environment than from studying animals that are closely related genetically but adapted to different environments behaviorally. Savanna-living social carnivores such as lions, hyenas, and hunting dogs could serve as valid behavioral analogues for early hominids, whereas the forest-living apes are not as useful. Schaller and Lowther deduced that early hominids would have been neither exclusive hunters nor exclusive scavengers. Like the living carnivores, they would have been opportunists. Sometimes they would have killed sick or wounded animals. Schaller and Lowther caught a sick antelope by the tail to simulate a kill by nonweapon-using early hominids. Early hominids would also have eaten whatever they could find easily accessible or already dead but still edible. Such food items as bird eggs on the ground or a slow-moving tortoise were fair game.

Hominids, like other social carnivorous mammals, would have had territories that they would have defended against competitors, mainly members of other groups of hominids. Territories function to spread the members of a species out over the terrain so that there are sufficient resources for it to survive. In species that live in social groups, like baboons or lions, troops or prides each have specific territories.

Robert Ardrey, a playwright schooled in anthropology at the University of Chicago, first popularized the concept of early hominid territory and territorial defense. His book entitled *The Territorial Imperative* appeared in 1966. It was one of the first popular works to apply purely ethological principles to early hominids in an attempt to explain what behaviorally sets us apart from our closest primate relatives. As he summarized some years later in *The Hunting Hypothesis* (1976), "the physical defense of a territory on the part of a hunting society meant a psychological union that the chimpanzee never knew."

## BEHAVIOR AND EARLY HOMINIDS

Ethologists and anthropologists reasoned that climatic changes during the Pliocene and Pleistocene epochs (the last two and a half million years) would have caused territories to increase or decrease in size. When droughts reduced the availability of food and water, groups of hominids would have come into conflict over those scant resources, just the same as social carnivores, like spotted hyenas or cheetahs, in recent times on the African savanna. Those with greater group solidarity would have survived and reproduced. Those who did not cooperate and successfully gain the resources necessary for survival would have lost out and become extinct.

Ardrey's early hominid realm was a world red in tooth and claw. Yet it was a world that he had reconstructed from the work of scientists and for which supporting data existed from the sciences of ethology, ecology, geology, paleontology, and anthropology. Ardrey's work is important to our story because it was the first widely popular and extensively read synthesis of paleoanthropology with ethological research.

My first anthropology instructor, Denny Salzmann at Phillips Exeter, gave me a copy of Ardrey's *African Genesis* (1961) to read in 1967. Ardrey took the plot for *African Genesis* largely from Raymond Dart, the discoverer of *Australopithecus*. It was a powerful collaboration: Ardrey, the playwright, and Dart, the theatrical paleoanthropologist. When Dart made a point it was with emotion. Sherry Washburn once told me that he had accompanied Dart on his rounds to the medical student dissection tables while visiting the paleoanthropology laboratory in Johannesburg. Dart was not pleased with the quality of the students' dissection and suddenly became quite emotional. As tears welled up in his eyes and ran down his cheeks he recounted how early anatomists had sacrificed for the privilege of dissecting the human body and how medical students had been forced to protect their anatomy professor from angry street mobs trying to lynch him for suspected grave robbing. And now, after all that, seeing how these students were squandering their hard-won opportunity made Dart feel quite overcome. As he walked away from the dissection table, now surrounded by extremely serious students, he turned to one of his assistants and said in a suddenly composed voice, wiping his face, "that should hold them for a while."

Dart's view of *Australopithecus* was of a "bloodthirsty, killer ape," to use his phraseology. Dart's ideas came from his interpretations of the fossils that had come out of the quarries of South African caves, bone assemblages that had included the bones of australopithecines. None of the South African early hominid fossils had intact skull bases. Dart took this to mean that the *foramen magnum*, the hole at the base of skull through which the spinal cord passes, had been artificially enlarged by the hominids to eat the brains of the deceased. A number of animal bones found at the sites had been smoothed to sharp points. Dart believed these to be australopithecine bone daggers. Long bones of antelopes appeared to have battered ends. Dart concluded that these were australopithecine clubs. A number of the fossil skulls of baboons, and some hominid skull remains, had depressed fractures that Dart believed matched the diameters of the ends of the bony clubs. He concluded that the preponderance of depressed fractures on the left side of the frontal bones of baboon skulls showed that they had been dispatched by right-handed club-wielding australopithecines. Dart went so far that even lack of evidence became evidence for the hypothesis. There were very few vertebrae from the tails of animals in the cave sites. Dart wondered why. He suggested that the tails of animals had been used by the australopithecines out on the savanna as antipersonnel weapons or for defense—whips. Thus the small tail bones would not be found in the bone assemblages in the caves, where he believed the hominids lived.

Robert Ardrey bought Dart's colorful view of the australopithecines and incorporated a killer-ape model for early hominids into *African Genesis*. It became widely accepted that humans are the only species that kills its own and that our nature is basically aggressive. The opening sequence in *2001: A Space Odyssey* was inspired by this view of human nature and was choreographed by Dart's star student, Phillip Tobias. Dart,

Tobias, *et al.* believed that we needed to accept the fact of human aggressiveness and then try to cope with it in a dangerous world, now proliferating with nuclear weapons. Indeed, one of Dart's crusades in his latter years became the prevention of nuclear war. At a conference in 1974 in San Francisco celebrating the fiftieth anniversary of his discovery of the first australopithecine, Dart, who was well into his eighties, was supposed to address the audience for fifteen minutes on the historical significance of the discovery. But after his first few sentences he launched into an hour-long, podium-pounding diatribe against nuclear war. Finally, the conference organizers managed to lead a slyly smiling Dart away from the podium, leaving in his wake a bewildered audience and a row of frustrated academics on the stage.

Following *African Genesis* a number of papers and books appeared assessing and reassessing the evidence and the conclusions regarding the nature of early hominid behavior. Paleoanthropologists and paleontologists looked anew at the fossil evidence from the South African caves. C. K. Brain of the Transvaal Museum in Pretoria did much of this pioneering work. It was discovered that most if not all of the bones that Dart had thought had been smoothed, broken, or battered by early hominids had been chewed up and partially digested by hyenas. The smoothing had been effected by hyenas' very strong digestive juices. The depressions in the skulls had been made not during life by australopithecines with clubs but after death by rock fragments in the cave sediment being pushed slowly over many years into the bone surface by geological pressure. And it was discovered that the hominids almost certainly did not live in these near-vertical caves. Most of the bones had been kills and had been dropped into the caves by leopards and saber-toothed lions lounging in or under trees at the cave entrance. There was even an alternative and more probable explanation for the lack of tail vertebrae. When leopards tear off the hide of a prey animal and discard it, the tail vertebrae stay with the hide, not with the rest of the carcass that the leopard may then carry to its tree lair. The tail vertebrae thus would not tend to fall into the cave with the rest of the skeleton when the leopard finished its meal.

A different picture of early hominid behavior and ecology began to emerge with these in-depth paleoanthropological reassessments. Rather than marauding killer apes, australopithecines became more like victims, living in a dangerous environment. C. K. Brain recognized a skull of a juvenile australopithecine at the South African cave site of Swartkrans with leopard tooth marks through its eye sockets and in the back of its head. A leopard had killed the hominid and then dragged the body, face forward, between its legs up into a tree over the cave entrance, from where the skeleton had fallen into the cave site. One of Washburn's early papers on the subject was entitled *"Australopithecus*: The Hunters or the Hunted?" (1957). Most anthropologists came to believe that early hominids were more of the latter than the former.

But these reassessments did not change the fact that what Dart and Ardrey had put their collective finger on was of the utmost theoretical and practical concern. Aggression, territoriality, murder, war, and crime were and still are major problems for all societies. If the paleoanthropological record could throw little new light on these issues, how else could we determine how human behavioral evolution had progressed?

Ethologists then came back into the picture. British zoologist Desmond Morris led the way. He took the paleoanthropological scene of early hominid evolution and painted in the behavioral details from animal behavior, particularly primate behavior.

Morris's first major popular work was *The Naked Ape*. He focused on the fact that human beings are largely hairless, unique among the primates. A thick coat of hair is important for most primates for the basic heat and insulation properties that it provides. But in most primate species hair also became the locus for one of the most important of all primate social functions—grooming. Primate behaviorists starting with Washburn and DeVore noted that primates spend an inordinate amount of their nonfeeding and nonsleeping time carefully picking through the hair of their fellow group members. They pick out any ticks, dirt, sticks, or other foreign material that they find. But the pleasure that this activity gives to the primate being groomed seems to transcend any utilitarian purposes. The expressions that primatologists observe in primates being groomed range from dreamy passivity to sleepy rapture. The grooming primate also seems to derive pleasure from the exercise, perhaps an emotion resembling human contentment. Females groom males after copulation, dominant males groom infants, subdominant males groom dominant males, juvenile males groom females, and so on. Grooming is a type of social glue that holds primate societies together. Or at least it is a manifestation of the social glue. What happens in a primate society, such as early hominids, when the hair that was groomed disappears?

Hominids lost the coat of hair that covered their bodies in response to natural selection. A hairless body allows sweat to evaporate easily and thus can cool off body temperature in a hot climate. Most of the surface area of our bodies is devoid of hair but well-supplied with sweat glands. This adaptation is unique among the primates. It allows hominids to undertake active lifestyles, such as hunting and gathering, in the heat of the day in hot tropical conditions. Hominids cannot, however, stray too far from fresh water, which is needed to replenish the body's water lost through sweating.

Morris hypothesized that hairless skin, as a sense organ, became very important in hominid evolution. Touching became the most basic of human communication devices. The greater sensitivity of human skin became another avenue whereby bonding could take place between members of human society, particularly in the male–female pair bond.

Morris took concepts from ethology and applied them to the human species. He postulated that just as Tinbergen had found certain environmental stimuli that had triggered courtship and mating behavior in birds, so also must there be such "innate releasing mechanisms" in humans. Morris thought that these cues were from both smell and sight.

Sexual cues that came from the sense of smell provided a key to the distribution of most of the rest of hominid body hair. In the pubic area and in the armpits there are sweat glands of a different type than occur over most of the rest of the body. These are known as apocrine glands and they secrete an oily substance that reacts with skin bacteria to produce the unique human smell that the deodorant industry makes millions each year for masking. The hair that grows in the armpits and the pubic area holds this scent. Very similar to musk, used by the perfume industry to stimulate olfactory interest in the opposite sex, normal human body odor is a natural aphrodisiac.

Since primates are fundamentally visual creatures, Morris suggested many more sexual cues based on visual stimuli. He started with the buttocks. Humans have large gluteal muscles to extend the leg fully in walking and to hold up the weight of the trunk against gravity in the erect posture. Morris hypothesized that the buttocks became in hominids what the periodically swollen sexual skin around the vagina is in chimps—a sexual releaser. Hominid males essentially were genetically programmed to respond with heightened sexual interest when presented with this stimulus. In keeping with the fact that hominid females did not have an estrus period, the stimulus was permanent. Coverings of the genitalia—clothes—may have evolved not to provide warmth in the hot tropics of our ancestors but to control the sexual responses of males.

But what of the change in copulatory position from back to front that, according to our scenario, accompanied the appearance of hominids on the scene? Morris had an answer. The hominid female evolved sexual releasing stimuli on the front of her body. The enlarged breasts of human females elicit the same sexual interest that buttocks had elicited from behind. Only now they cemented the male–female bond that had become the hallmark of hominid social structure. This type of adaptation is not unknown among nonhuman primates. The bleeding heart baboon, the gelada of northern Ethiopia, has a bare patch of skin on its chest. During estrus this skin in females becomes very noticeably red, signaling that the females are ready for copulation. Geladas have evolved this trait because they spend much of their feeding time during the day sitting and "bottom shuffling" from one source of food to another. Their pudendal areas are thus seldom seen. Yet once sexual interest is aroused in gelada males, sexual mounting is from behind.

The wave of popular and scientific interest in the evolution of human behavior fostered by ethology had a lasting effect on the field of human evolutionary studies. Like the population biology and renewed commitment to biological theory that resulted from the Cold Spring Harbor conference, ethology represented another biological discipline that injected its theoretical perspective into anthropology. Study of primate behavior in the wild, not under captive conditions as psychologists were wont to do, and the interpretation of behavioral patterns in evolutionary terms, made observations of nonhuman primates relevant or potentially relevant to human evolution. Speculations about early human behavior abounded, but in keeping with the experimental nature of the new physical anthropology, behavioral scientists now had a paradigm to use to test models of primate behavior. And paleoanthropologists had models that they could test with the fossil and archaeological record.

## RESISTANCE TO ETHOLOGY

Morris's naked ape hypotheses and the new wave of ethological theorizing about the evolution of human behavior did not sit well with a number of anthropologists, particularly the breed known as sociocultural anthropologists. To suggest that human beings acted out behavioral patterns that they had inherited from their ancestors and that had been encoded in their genes was abhorrent to them. The whole approach was condemned as "biological determinism" and equated with racism. Anthropological debate sometimes can take on a testiness, prickliness, and even urgency that does not typify the other sciences. Perhaps this is because the issues are so close to us and of such immediate practical concern.

Sociocultural anthropologists instead believed on the whole that human beings learned their behavioral patterns through that complex social system of patterned learned behavior known as culture. Although not quite as naive as the romantic philosopher Jean-Jacques Rousseau's notion of the human mind being a blank slate at birth, a tabula rasa, the position of the sociocultural anthropologists considered the biological, inherited, hard-wired aspects of human behavior to be relatively unimportant. Some far distant primate heritage might underpin human aggression in all human societies, for example, but the sociocultural components causing the behavior were far more important. Sociocultural anthropologists might suggest that to reduce aggression in modern American society one must change the underlying social message that reinforces the behavior—perhaps by removing plastic assault rifles from children's toy stores, getting violent cartoons off television, and educating members of society about the value of individual, ethnic, and cultural differences. While many biological anthropologists might think that these goals are laudable, few of us would believe that this solution would get to the root of the problem.

Most of us biological anthropologists, like our cousins from human ethology consider aggression, like much behavior, to be to a large extent hard-wired, not programmed, into the human animal. The behavior evolved to serve very important survival needs over the very long time span of our species and of our species' ancestors, and it is still there in us. When food or water were scarce our ancestors fought successfully for access to that food and water, not only for themselves but for their families, relatives, and group members. The group excitedly cheered them on—"push 'em back, push 'em back, *way* back," "two more yards, two more yards," "hold that line"—and when they succeeded the group roared its approval. A football game is an allegory of the human territorial and aggressive adaptation. Ethologist Konrad Lorenz in *On Aggression* suggested that national sports are perhaps the best way to deal with what he considered an innate human predisposition to territorial defense and vigorous physical competition. The World Cup and the Olympics seem far better alternatives than nuclear war.

## THE PARADIGMS OF CULTURE AND EVOLUTION

The fundamental difference between the sociocultural anthropologists and the ethologists revolves around their paradigms—those templates of thought that organize our interpretations of the world around us and that organize our behavior based on those deductions. The paradigm for the sociocultural anthropologists is culture. The paradigm for the ethologists is evolution.

Culture is what separates human beings from animals. It is the sum total of everything we know as a species. Children are bequeathed the particular manifestation of culture into which they are born—the language, customs, and outlook of the group, termed "a culture"—through the process of *enculturation*. But a member of any particular culture can learn all the nuances of any other culture, given enough time and if one starts young enough. A transplanted Maasai tribesman from Kenya can grow up in England, attend Eton and Oxford, and come away with as much an upper-class English cultural background as Prince Charles. Culture is interchangeable. This interchangeable universal nature of culture, clearly unrelated to racial or biological heritage, made sociocultural anthropologists believe that human behavior has little if anything to do with biology.

Several discoveries have changed this view, at least in the eyes of biological anthropologists. The definition of cultural behavior in the 1960s included such quintessentially human attributes as tool-making, the ability to symbolize, language ability, and behavioral differences among different groups. But then Jane Goodall and her ilk came along. They discovered that chimpanzees make tools. They tear off twigs from plants, pull off the side leaves, and carefully insert these tools into termite nests, where the termites attack the stick as an intruder. The chimps then methodically extract the tool and lick off the insects for a high-protein lunch. Another chimp researcher, Bill McGrew, found that chimp groups have different local customs. And numerous laboratory studies of chimps began to demonstrate that they can use symbols to identify concepts. Chimps can be taught that a red plastic triangle, for example, might mean water. And chimps such "Nim Chimsky" learned American sign language, the hand language used by the hearing impaired, although with a reduced vocabulary. Clearly, human biology did not determine all the behavior in these animals. They could learn a significant amount that had in the past been considered human. Although the ability to learn might seem to support sociocultural anthropological claims, Goodall's discoveries reversed them.

When sociocultural anthropologists had pondered how culture had come into being, which admittedly was not too often, they thought that it had been at a critical point, something like the boiling point of water. Heat builds up slowly, but only at 100° C (212° F) does water boil. There are no intermediate steps. One either has boiling water or nonboiling water. The chimp data caused real problems. How could there be half culture? Most sociocultural anthropologists ignored the problem.

Then, archaeologists began looking at increasingly earlier time periods in order to discover the origins of stone tool making, a convenient marker for the beginning of culture. The early phases of stone tool making, the crude flakes and hand-axes that *Homo habilis* and *Homo erectus* had used, were very simple. But interestingly, these stone tools, which first appeared at around 2.5 million years and 1.5 million years ago, respectively, changed very, very slowly, if at all, over hundreds of thousands of years. They were not improved or modified by the cultures that used them for immense periods of time. Coming from a perspective of the modern world in which even last year's ties are out of fashion, such cultural conservatism is almost unimaginable. There was something distinctly subhuman about culture in its initial stages. Archaeology reinforced the conclusion from primate behavior studies that culture had not arisen at some critical point but had gone through a number of intermediate stages.

In other words, culture had evolved. The fact that it had evolved became a defining characteristic of human culture, as distinct from the patterns of social behavior of other primates. But in opposition to sociocultural anthropologists' views, this cultural evolution did not offer a "boiling point" for humanity.

The sociocultural anthropologists then were left with a cultural research paradigm that limits them to looking at only modern humans in a noncomparative way. This is a serious drawback because science in general makes progress when discoveries in one area can be applied as hypotheses to be tested in another. If one's paradigm rules out the comparison, then one is left with few if any hypotheses to test. Sociocultural anthropologists can and do undertake comparative studies of different human groups, but as cultural homogenization and global intercommunication have accelerated even these comparative studies have become confused.

On the other hand, biological anthropologists, including the primate behaviorists, ethologists, and paleoanthropologists, switched paradigms, if indeed they ever were fully in the cultural paradigm fold. They now orient their research around evolution, using natural selection as the process by which they test hypotheses. But the years of association with the sociocultural anthropologists have emphasized how complex and important modern cultural adaptation is. It just cannot be understood sui generis. Biological anthropologists are interested more in the species-wide nature of culture as an adaptation, not the differences between individual cultures. They are thus concerned more with the biological and genetic roots of behavior.

This clash of paradigms has set up strong centrifugal forces in academic departments of anthropology, where sociocultural anthropologists far outnumber their biological and archaeological colleagues. Sociocultural anthropologists find themselves a bit like geographers—a group with important information to impart to society, but without a research frontier to explore. Like the geographic frontiers of the earth, which are all well-mapped and explored, all the cultures of the world either have been studied and written up or they have disappeared, swallowed in a swirl of westernization. Will the sociocultural anthropologists follow their colleagues in changing their paradigm and adopting a comparative approach (which, after all, is the basis of anthropology) or will they persist in holding onto a paradigm that now holds no predictive value? The answer probably is the latter.

But for biological anthropology the future is now. The late 1960s saw the launching of the largest and best-funded research project into human origins up to that time, the

International Omo Research Expedition to southern Ethiopia. The scientific program of the expedition, directed by F. Clark Howell of the University of California at Berkeley, and codirected by Louis and Richard Leakey of Kenya, and Camille Arambourg and Yves Coppens of France, included a multidisciplinary team that gathered data of an unprecedented completeness covering a period of time spanning over 3 million years, from the early australopithecines to the appearance of *Homo sapiens*. For the first time major evolutionary transitions in the human lineage could be dated at one place and in one long geological sequence, and the environmental, ecological, and cultural attributes associated with them could be investigated directly. Evolutionary and behavioral hypotheses could be tested with data, not debated endlessly. The Omo expedition represented a mature application of the new physical anthropology approach in paleoanthropology. This expedition and its findings are the subject of the next chapter.

# 3

# At the Navel of the World

Nairobi, Kenya, May 1973. Although Kenya had become independent nine years before, in 1973 it was still to my eyes an outpost of the British Empire. My great grandfather had come from England, and I had been regaled from earliest childhood with Kipling stories and general Victoriana. Urban Kenya was up to that time the closest I had come to England and I found all the cultural details and turns of phrases fascinating.

But I was also somewhat disconcerted, as I sat alone in the quaint little Ainsworth Hotel dining room in Nairobi the day after my arrival. No one was there to meet me. I had understood that Dr. Frank Brown, a geologist recently finished at Berkeley and now at the University of Utah, would be in the hotel and would meet me when I got in. He was nowhere to be seen, and the hotel desk did not have him listed as a guest. So I just waited. At ten o'clock that night there was a knock at my door. It was Frank Brown, covered in oil, and he wanted to know if I would help rescue a broken-down Landrover. My African explorations had begun.

## AFRICA AS THE BIRTHPLACE OF THE HOMINIDS

My first intellectual ramblings into human evolution were not geographically focused. As a sixth grader I was only vaguely aware that the cavemen I had read about had lived in Europe, and not in North America, where I lived. I tended to think of them in the same way that dinosaurs had been depicted in one of my early books—as translucent ghosts standing next to your house. It took me even longer to grasp that the very earliest ancestors of people had lived in Africa. Indeed, this was not generally agreed upon in scientific circles.

I did not seriously begin to apply myself to learning about Africa until my last year in undergraduate school, when it became apparent that I was going to have to go there if I wanted to study early hominids.

Africa was traditionally thought of as a backwater of human evolution, if it was considered at all. The German anthropologist Hermann Klaatsch, for example, had proposed in 1910 that modern humankind had arisen in Asia, a spin-off of orangutan stock, whereas the more primitive Neanderthals had come originally from Africa and were derived from a gorilla or chimpanzee ancestry. No fossil evidence supported this fanciful notion. It now seems that Klaatsch had it exactly backward. Neanderthals seem to have been a local evolutionary product developed in Europe, and anatomically modern people, as indicated by the fossil record, seem to have appeared first in Africa.

Our knowledge of African origins has not been won easily. This chapter explores how theory and science have gradually altered our understanding of a key element in the origin search: where did the missing link live?

Through much of the nineteenth century Asia was the more commonly hypothesized continent of origin for hominids. Both Alfred Russel Wallace, the codiscoverer of evolution by natural selection with Darwin, and Ernst Haeckel, Germany's first great evolutionist, preferred an Asian origin theory. Eugene Dubois, the Dutch physician and paleoanthropologist, had gone to Java looking for the missing link in the 1890s and had discovered *Homo erectus*. Isolated teeth of an early hominid had been found at the cave of Zhoukoudian near Beijing in northern China by Austrian and Canadian paleontologists. It was reasonable to think that the ancestor of *Homo erectus* in Asia could have been the true common ancestor with the non-human primates that evolutionary theory predicted.

In 1922 Henry Fairfield Osborn, the great American paleontologist, friend of Teddy Roosevelt, and President of the American Museum of Natural History, sent from New York to Asia the most lavishly funded paleoanthropological expedition ever mounted. Its express goal was to discover the "missing link." Known as the Central Asiatic Expedition, with American flags flying and with all the scientific members wearing matching expedition sweaters, it set off inland from the Chinese coast with twenty-two trucks and over one-hundred camels. As the expedition wended its way through China to the Gobi Desert, stopping to collect specimens and excavate along the way, a constant train of camels ferried gasoline and other supplies from the coast. Roy Chapman Andrews, the expedition's field director, estimated that it cost over $1 million (in 1924 dollars) to launch the Central Asiatic Expedition and to maintain it in the field for three years. It was the most elaborate of Henry Fairfield Osborn's paleontological expeditions, already known for impressive results. Osborn's collecting expeditions had been so successful, for example, that his was the only museum able to mount a whole series of original fossil skeletons depicting elephant evolution. The simple dramatic force of the skeletons walking through time is so eloquent that the exhibit has stood unaltered for over half a century in the Hall of Mammals at the American Museum. But the Central Asiatic Expedition was to be Osborn's tour de force, an all-out push to discover the ultimate mystery—our human origins—and a vindication of Osborn's theory of Asian origins.

The rationale that impelled the Central Asiatic Expedition into the depths of Asia was Osborn's hypothesis that hominids had arisen as a part of the vast modern faunas of Holarctica, the northern hemisphere of the Old and New Worlds, along with the horse, the elephants, and many of the carnivores.

Although Osborn's idea on the place of origin of the hominids found general favor among other scientists, he also held the idiosyncratic view that apes had had nothing to do with hominid ancestry. He believed that the large human thumb was so unlike the relatively reduced ape thumb that hominids must have derived from "less specialized" primates much earlier in evolutionary history than the apes. He hypothesized a premonkey point of divergence for the hominids, in Asia, early in the Cenozoic geological period, much earlier than the cave people discovered at Zhoukoudian.

The long train of trucks and camels snaked its way across the Asian steppes, stopping to investigate geological outcrops for fossils and concentrating on those that were much earlier than the Pleistocene deposits that had begun to yield Peking Man. Fantastic paleontological discoveries were made. Vast numbers of fossil antelope and carnivore skulls from the Miocene Epoch were found, which the expedition's technicians encased in heavy plaster jackets, and then sent back by camel to the coast for eventual shipment to New York. Dinosaurs were discovered, including the first dinosaur eggs. But after three years of massive effort, no hominid—indeed, no primate—fossils had been discovered. It was apparent that nothing in the tons of fossils that had been collected in central Asia supported Osborn's hypothesis.

In 1925, as the Central Asiatic Expedition was winding down and heading back to New York, a strong claim for human origins on the African continent was made. Raymond Dart's landmark discovery at Taung, South Africa, of the juvenile fossil hominid skull that he named *Australopithecus africanus* was published in *Nature*, the British scientific journal. The paper generated controversy, especially among the British anthropological establishment, who in general had very firm ideas about how the human species had originated. Indeed, a number of them had been knighted by the crown for their contributions. Their hypotheses did not include Africa.

Sir Arthur Keith, an anatomist of formidable reputation at the Royal College of Surgeons, criticized Dart's claims for a close association of the new skull with hominids. It was likely a juvenile chimpanzee, Keith said. Such a mistake, he magnanimously noted, could be made because the skulls of juveniles of apes and humans are much more similar than are the adult skulls. Keith had no particular reason to go easy on Dart, a still wet-behind-the-ears former student of one of his professional rivals, Sir Grafton Elliott Smith, and now in some far-flung corner of the empire.

But the 1920s and 1930s were a period of theoretical upheaval in paleoanthropology. Sir Arthur's dismissal of Dart's claims, which might at another time have sounded the death knell for the international career of the upstart anatomist from South Africa, only made Dart redouble his efforts. Dart ensured that paleoanthropologists who had been secure in their hypotheses of non-African, northern hemisphere hominid origins were haunted by the vacant, staring eye sockets of the Taung baby. He went so far as to suggest that world maps should be redrawn. Other than the presumption of northerners, he said, there was no reason that north was put at the top of the world. He preferred a globe with South Africa and his homeland, Australia, in the upper part. The hominids then came from "up south" and not from "down north."

The world was not quite ready for such a radical switch in thinking. Other alternatives were out there. The Asian hypothesis was still supported by Osborn, and important discoveries were being made by a Canadian paleoanthropologist named Davidson Black at the cave of Zhoukoudian in China. For a very short time hope existed that hominid origins might be traced to North America. In 1928 a fossil tooth of a peccary, an American pig-like mammal, was found in Nebraska by one Harold Cooke and mistakenly named as a new species of ape, *Hesperopithecus haroldcookei*. The paleoanthropologist William King Gregory, who otherwise had a stellar career at the American Museum, later realized his mistake and corrected it.

Keith and many of his colleagues believed that Europe was the ancestral homeland of the hominids. Keith's support came from evidence from the gravel pit at Piltdown, in Sussex, England—evidence that some other anthropologists regarded as suspect at the time and that proved fraudulent in 1953.

Piltdown Man was a chimera. A modern human skull had been broken and stained to look ancient, and a chimpanzee jaw had been broken and the teeth filed flat to appear human-like. The pieces had been placed by persons still unknown in the gravel at Piltdown for scientists to find. In the hands of Keith, the master anatomist, the reconstruction had been simple and the conclusion straightforward. The earliest presumed human fossil then known had had a large brain and an ape-like dentition, with a large canine tooth. Dart's *Australopithecus*, on the other hand, had a small brain and a small canine tooth. From Keith's standpoint not only was Dart's discovery in the wrong place, it had all the wrong anatomy.

Whether Piltdown was a cruel joke perpetrated on Sir Arthur Keith or, as Frank Spencer has recently hypothesized, a cruel joke perpetrated by Keith on an unsuspecting scientific community may never be known. But Keith's reputation certainly suffered as the discoveries of australopithecines from Africa mounted through the 1940s and 1950s. Nothing further came from Piltdown, or anywhere else for that matter, that supported Keith's position. Finally, as new chemical dating techniques came into vogue in the early 1950s, Piltdown was shown to be modern and thus fraudulent. Confronted with the news in his waning years of retirement, Keith gave up his hypothesis of a Eurocentric origin of hominids with large brains and ape-like teeth. The hypothesis had become termed the "pre-sapiens hypothesis" because it posited that large-brained hominids had existed long into the past and that other, less advanced hominids, such as Neanderthals, had become extinct side branches.

Louis Leakey, who had studied at Cambridge during the heyday of Piltdown in the 1920s, strangely enough was to become the torchbearer of a new version of the pre-sapiens hypothesis. Leakey, of course, changed the venue from Europe to Africa. But the idea of a long-lived, large-brained human lineage, distinct from more primitive contemporary hominids, persisted. Leakey was searching for fossils to support the hypothesis as early as 1931 on the savannas of East Africa. He finally found evidence in 1961 at Olduvai in the form of *Homo habilis*. Raymond Dart was all too happy to suggest the new name.

## RESEARCH BEGINS AT OMO

A few years after the sensational discoveries at Olduvai, Leakey had the occasion to meet Ethiopian Emperor Haile Selassie at a state dinner in Nairobi. Selassie asked if sites like Olduvai existed in his country. Without hesitation Leakey mentioned Omo. The emperor became interested and encouraged Leakey to come to Addis Ababa to make arrangements for an expedition under his direct imperial authorization. Leakey contacted Camille Arambourg in France and Clark Howell in the United States, who both had worked at Omo, and the International Omo Research Expedition, with Kenyan, French, and American contingents, was organized.

Clark Howell, informed by the new biological anthropology's need for context, wanted to avoid some of the worst mistakes made by earlier expeditions. In particular he wanted to get a reliable geological framework for the fossils that the Omo expedition was to collect. Without that information, paleontologists could not tell how the various collection sites in the vast area at Omo might relate to one another within the layer-cake history, or stratigraphy, of the region. Interminable debates would then ensue over the exact age of the finds and the relationships of some fossils to others. Many times these debates at other sites were unresolvable because the fossils had been collected years previously without clear geological or even exact locality contexts. Howell wanted to ensure that the Omo research got off on the right foot and that the geology was done before, not after, the first fossils were collected.

He walked across campus at Berkeley to Garniss Curtis's lab in the Earth Sciences Building. As he was talking with Curtis about the need to find a geologist willing to go out on a solo geological survey to a very remote location in southern Ethiopia, a voice behind them suddenly volunteered. Both Howell and Curtis turned around. The speaker was Frank Brown, a graduate student working in Curtis's lab on mineral

separations for his doctoral dissertation on Ugandan volcanoes.

In June 1966 Brown found himself headed north from
Nairobi, Kenya, into what was known as the NFD—the
Northern Frontier District. He drove a Landrover pulling a
small water trailer. His route took him west of Lake Rudolf,
into a little corner of Uganda, and eventually to the border
post of Todenyang. Past Todenyang the road into Ethiopia
became difficult to spot—two tire tracks in the flat rift valley
floor, obscured by the hoofprints of hundreds of cattle. He
looked for the Ethiopian border post of Namuraputh and eventually found it perched atop a wind-swept massif. The soldiers
were a bit scruffy and wild-eyed, as you might expect of men
posted to this forlorn and desolate place. After being cleared
through Namuraputh he pressed northward into the Omo
River valley and reached the general area after a week's travel
from Nairobi. At the police post of Kalam, he hired a local
member of the Dasenetch tribe named Atiko, who spoke
Swahili and could serve as a guide, interpreter, and field assistant. Atiko was to work for the Omo Expedition each summer
for the next eight years; using his salary to invest in cattle and
goats, he eventually became one of the wealthiest men among
the Dasenetch.

Brown began to explore the region with Atiko. Near the village of Shungura he found terrain that looked like the badlands of the American West. This area had beautifully exposed
geological deposits, faulted and turned up on their sides at
about a twenty-degree angle. He saw tuffs, those grey horizons
of ash left over by ancient volcanoes, so important for potassium-argon dating. The tuffs were sandwiched in between yel-
low-brown and reddish-brown layers of rock laid down by lake
and river waters. The rocks exposed here were to give their
name to all of the geological units in the main part of the Omo
Valley—the Shungura Formation. Brown established the roads
by which the expedition members reached their field camps the
next year and by which they negotiated their way down to the
Omo River.

With Brown's pioneering efforts to go on, the American,

French, and Kenyan contingents of the Omo Research Expedition launched themselves into the field during the summer of 1967. Professor Clark Howell headed the American team; Professor Camille Arambourg of Paris led the French team; and neophyte Richard Leakey, standing in for his ailing father, undertook his first real paleoanthropological expedition as the leader of the Kenyan contingent.

Richard Leakey thought that the French and American teams, headed by two senior professors, had made a pact between them to divide up the oldest and largest two sectors of the Omo deposits, respectively. He and the Kenyan team got what was left over—a small and geologically quite young deposit, known as the Kibish Formation. But as he did during much of his later career, Richard Leakey turned a potentially bad situation into an advantage. He discovered two quite complete human skulls, anatomically modern *Homo sapiens* to be sure but nevertheless dated to the very early time of 120,000 years ago. They are to this day the oldest record of modern *Homo sapiens* in the world. The discovery eclipsed the many well-preserved, but largely nonhominid, fossils found by the American and French teams, dated to several millions of years. The French even found the jaw of an early hominid, which they had named in accordance with the old typological principles by then largely abandoned elsewhere, *Paraustralopithecus aethiopicus*—a new genus and species. But most of the attention was garnered by Richard Leakey's spectacular human skulls.

Richard did not stay around the Omo to savor his success. Borrowing the helicopter that the expedition had chartered to get the geologists around the deposits, he flew across the Ethiopian-Kenyan border and landed on the eastern shore of Lake Rudolf. If the professors would not let him and his team look in the old deposits at Omo, he intended to find some of his own. He did exactly that. On one of his landings Leakey found both fossil mammal bones and primitive stone tools. He knew he had located a previously unknown site of some potential significance.

When Louis Leakey was preparing to travel back to the United States for his annual lecture tour, Richard surprised his father by asking if he could come along. Richard previously had never shown any great interest in the business of fossil-hunting and fund-raising. But his father agreed. When they arrived in Washington, Richard had another surprise for his father as well as for the National Geographic Society Committee on Research and Exploration, whom his father had come to address. He boldly asked for funding to exploit the important site that he had just discovered east of Lake Rudolf. The committee was put off by Richard's brashness, but decided to grant him the money. They told him, however, never to come knocking at their door again if he failed.

When Richard Leakey arrived back in Kenya and had made his way back up to Lake Rudolf, he was embarrassed to discover that he could not find the site that he had originally located with the Omo expedition helicopter. But the concern did not last long. The low hills east of the lake might be unimpressive scenery, but the ancient lake beds preserved fossils in a state of completeness rarely if ever encountered at Omo. By 1972 Richard and his team had discovered complete limb bones and nearly whole skulls of early hominids, whereas the Omo professors had only fragmentary teeth and jaws. The world was electrified by his finds. Richard Leakey himself was electrified by the rapidity of his success. He published initial reports on his finds and started on a lecture tour of the United States and Great Britain.

I first met Richard Leakey where I had met his father a year before, after his lecture at George Washington University's Lisner Auditorium in Washington. Despite the general bad publicity then circulating in academic circles to the effect that Richard had never had any formal university or graduate school training, he gave an impressive performance. Articulate and smooth in his delivery, he succeeded in conveying the excitement of the discoveries. He veered adroitly away from areas too technical. After the lecture I asked him about his statement that one of his fossils, a humerus—the upper arm bone—showed that its possessor in life had been bipedal. "Since the humerus is not involved in walking, except passively swinging as we take one step and then another, how can you tell that its anatomy indicates bipedalism?" I asked. "By the general similarity in form to modern man, the straightness of the shaft, and the attachment areas for the pectoral muscles," he replied evenly. I was pretty sure that this reply didn't make sense from the standpoint of functional anatomy, and I was quite sure that only one pectoral muscle, *pectoralis major*, attaches to the humerus. But it was clear that Richard Leakey was a smooth operator, a fact that I was to appreciate much more fully in the years to come.

## RESEARCH IN SOUTHERN ETHIOPIA

Paleoanthropological fieldwork is a curious mixture of Boy Scout campout, East African safari, heavy construction work in the hot sun, and meticulous excavation that has always reminded me of anatomical dissection. These were my impressions of my first field season at Omo, a three-month period during the summer of 1973 in which I was introduced to African fieldwork. I was anxious to learn how early hominids were extracted from the earth and I was to find several that summer. But what struck me most was that after a very short period of time in the field, all the vague allusions that I had read for years about fieldwork suddenly became quite understandable because I had now experienced them.

Clark Howell called me during my last undergraduate semester and asked if I could go to the field early in order to assist in setting up the field camp before the other members of the team, including himself, were to arrive. I immediately agreed, even though I would miss graduation. The university would send my diploma to my mother.

The trip was a whirlwind. I met Frank Brown in Nairobi that night in May 1973, covered in oil. He asked if I could drive a Landrover. Yes, I lied. But it was just like driving a straight-drive pick-up truck. Luckily I didn't need to worry about the four-wheel drive in the city. On Brown's instructions, I immediately scurried around Nairobi to buy sacks of rice, get two-way radios repaired, get insurance papers in order, repair and clean the tents, patch Landrover tires, meet the official Ethiopian member of the expedition at the airport, and so on. Finally, we set off in a convoy of two trucks ("lorries"), loaded to their tops, and five Landrovers, filled with our thirty Kenyan workers. The only scientific members of the expedition on the land trek to Omo were Frank Brown, myself, and Harry and Joan Merrick, archaeologists. All of the other members of the expedition were to fly in once we had arrived and set up camp.

We went northwest from Nairobi on paved roads, descending into the great African Rift Valley down an impressive escarpment near Nakuru. The second day we turned due north, onto unpaved roads. After a while we stopped to let all the vehicles catch up. I walked for the first time on the African savanna. After the continuous drone of the Landrover engine was silenced, everything was very quiet. I listened to the sounds: a far-off bird with that insistent but laconic call that I had heard on nature documentaries, a buzzing fly, but nothing else. I looked closely at the short little acacia trees—their leaves and bark—the grass, the dirt. These were the sights and sounds that early hominids, my most ancient ancestors, had seen and heard. It was a hot, cloudless day, but the sun felt good. I had never been here before, but somehow the landscape felt familiar. I had the strange sensation that I had come home, for the first time since my ancestors had wandered away hundreds of thousands of years before.

When we arrived at Omo, we set up camp, cleared the airstrip, fixed the road to the river, and got water, which we pumped up from the Omo into forty-gallon drums. My first few days in camp were memorable. A dust devil, one of those swirling mini-tornadoes, blew in from the badlands to the east and hit my tent. I was in the dining area, and I watched as all my papers lazily spiraled several hundred feet into the sky. The Africans immediately helped me and fanned out across the savanna to pick up the debris as it fell to earth. A picture of my sister was retrieved several miles out; my notes on the African rift valleys drifted over into the Omo beds.

The first night my shoes disappeared. I had put my topsider boat shoes, one of those standard preppie accoutrements, immediately outside my tent flap. I couldn't find them in the morning. Frank Brown apologized. He said that he should have told me not to leave anything made of leather outside or a hyena would eat it. Later in the day one of the men found the sole just outside camp. Every bit of the leather upper had been chewed off. My family had had large dogs, usually collies, when we were growing up, and occasionally shoes or slippers would get chewed on. But this shoe was not merely chewed, it was *gone*. It had been eaten by an animal that meant business. I wondered what would have happened to me if I had not been protected by the tent. I made a mental note never to sleep in the open on the African savanna if I could avoid it. Hyenas reputedly go for your face first.

I was anxious to get started on excavating. Our goal was to find fossils of early hominids and the animals associated with them.

How do you figure out where to dig? Although it sounds like a tautology, the best place to find more fossils is where you had found them in the past. Harry and Joan Merrick started their archaeological site at Locality 396, where they earlier had found stone tools eroding out of the sediments. All the fossil and archaeological sites at Omo were assigned locality numbers and recorded precisely on aerial photographs. The geologists on the project went to the locality, measured and recorded its strata, and placed it in the overall stratigraphic layer cake of the Omo deposits named the Shungura Formation and divided into subunits called members. Locality 396 was geologically situated in Member F of the Shungura Formation, somewhat over 2 million years old by potassium-argon dating on tuff at the top of the member known as Tuff F.

Several paleontological sites had yielded up fossils and needed excavating. Frank showed me the one that he thought would be the best to start with, Locality 345, a hill north of camp and a distance down from the plateau into the Omo deposits. It was substantially older than Harry and Joan's site. Locality 345 was in Member C, more than 2.5 million years old. It is always difficult to know whether an excavation will be productive, but the best guide is whether fossils can be seen on the surface, exposed by water eroding them out of the enclosing sediment.

If the exact level from which the bones originated can be determined, an excavation may be successful in uncovering the fossils before they are broken up by erosion. Excavators can also record their original positions when fossilized.

Locality 345 had been partially excavated by another of Howell's graduate students, Donald Johanson, who was now writing his thesis on chimpanzee teeth in Chicago. The earlier excavation had revealed two partial fossil hippopotamus skeletons and numerous fragmentary bones and teeth of other animals. There was no indication that the bone level was petering out, but Johanson had stopped because the amount of overlying sediment, known as overburden, became progressively thicker as you went farther into the hill. If we were going to excavate Locality 345 we were going to have to remove some seven to eight meters of overburden.

The fact that the hippopotamus skeletons had been found with their various parts partially articulated indicated that at this site the conditions under which the bones had been buried had been relatively calm. The water that had laid down the sediments at Locality 345 had been slow-moving water, possibly a sluggish side-channel or back swamp of the Omo River. We could expect well-preserved hominids at Locality 345, if we were to find them at all.

I inherited Johanson's excavation crew. We started in with the gasoline powered cobra jackhammer, prying out blocks of sediment and shoveling and pushing them down the hill. Everyone hated their turn at the jackhammer. It was hot, dusty, noisy, and very tiring on the back. I took my turn with the Kenyans. This surprised them because *wazungu* (white people) rarely if ever touched the jackhammer. My philosophy, developed during my summers working construction during college, was that supervisors should pitch in and help whenever possible. And I was about twice as big as any of the Kenyans. We developed a good rapport, with my excavation team teaching me Swahili during the three months we spent digging together.

We started finding very nice fossils at Locality 345—teeth of antelopes, pig jaws, hippo limb bones, monkey teeth—all jumbled together in what was once an ancient back swamp fed by the Omo River. I started out slowly. This was a new medium for me. We were digging in rock—sandstone cemented with silt—with rock hammers and microchisels. These were large nails with flattened spatula-like ends that bit into a small bit of sediment and pried it loose. The excavation during the day sounded like an old-time tinker's shop, with constant pinging and hammering.

About three weeks into the excavation of Locality 345, one of the younger members of the crew, Patrick, called Mulali, generally considered the most accomplished excavator. Two or three others began to gather around, and then I went over. I heard the word *mtu*, Swahili for man. Could Patrick have hit a hominid?

What I saw was the rounded shape of a skull with a ridge of bone along the top known as the sagittal crest. It was just appearing from the enclosing sediment, and only the back top of the skull was visible. Not enough was exposed to tell what kind of animal it had belonged to, so I told Patrick to keep going, but very carefully. He had accidentally put a nick in the skull when he had first hit it. After the first day I was convinced the skull was a large primate, but the face, if there was a face, had to be exposed before we would know if it was a hominid. The next day and the next we slowly worked to remove the sediment from in front of the skull. The process was very much like anatomical dissection, except that what we were peeling off from the bone were layers of silt and sand and not the skin and muscles, which had disappeared millions of years ago.

The face was there. We descended down the bridge of the nose, down around the rims of the eye sockets. But the slope of the face began to protrude, unlike a hominid. As we went lower, we found that the face became elongated into a long dog-like snout. It was a large male baboon! When we reached the teeth we confirmed the identification. Clark Howell came out to the site and identified the baboon as *Theropithecus*, a relative of the living gelada of the northern Ethiopian highlands. He was quite excited. It was the most complete and undistorted skull known of *Theropithecus brumpti*, an extinct giant species first discovered by Arambourg at Omo in the

1930s. I was pleased that our excavation had been successful in producing an important specimen of a primate, but I was secretly disappointed that it had not turned out to be a hominid.

After five weeks we reached the end of the excavation area that we had carved out of the Locality 345 hill. I had carefully mapped and measured all the specimens, written specimen numbers on them (the baboon skull became L345-145), and wrapped them for shipment back to the lab. Should we remove more overburden and continue at this locality or should we try another spot? Clark Howell thought that we should spend the remainder of the season at another locality, one that he had intended to be finished by Don Johanson, who was still tied up with revising his thesis in Chicago.

We moved our excavation crew to Locality 398, a "bone bed" deposit laid down at the bend of the ancient Omo River. All the specimens here had been rolled along the stream bed and were broken and smoothed by the sand. But the sheer number of bones was unbelievable. At Locality 345 you could easily walk between the bones as they sat on their little pedestals of sediment in the excavation. When work at Locality 398 got underway, on the other hand, it was impossible to put your foot down anywhere without stepping on at least five specimens. Locality 398 was almost within sight of and at the same geological level as Harry and Joan's archaeological site, Locality 396. It was in Member F and dated to 2.1 million years ago. I thought that with all this bone it would be very likely the place that I would find my first early hominid, and I was right.

In my anthropocentric naiveté I thought that I would easily be able to recognize a hominid fossil. In my mind hominids had taken on such an aura that I was sure that they would somehow appear totally different from other fossils. But of course hominid fossils do not glow in the dark or have a little halo around them that identifies them to the investigator. When the first hominid appeared at Locality 398 it was surprisingly similar to all the other fossils—mute testimony that people are a part of nature, just the same as all the other species preserved at the site. The find was a flat-topped tooth that had only a part of the crown and that lacked the roots. At first sight it could be confused with a pig tooth; pigs, like hominids, grind their food with side-to-side movements of their jaws and have flat molar teeth for this purpose. But this tooth had a thick surface of enamel on its top, a characteristic of hominids. Thick enamel probably evolved to resist the forces of biting down hard on small, tough food items in the savanna. Forest-living African apes, who eat soft-textured fruits and succulent plants, have thin enamel on their molar teeth.

I pondered the scientific use of a tiny hominid fossil tooth fragment the size of your fingernail. It certainly was not going to add significant new information to the knowledge of early hominid anatomy. I suddenly lost my enthusiasm for the excavation. I told my mapping assistant, Kaumbulu, that I wanted to go and excavate at east Lake Rudolf, where they were finding more complete fossils. The long, hot days on the site became longer and hotter, and I began to look forward to getting back to the cool veranda of my tent and reading my Faulkner novel in the afternoons.

## THE DISCOVERY OF CONTEXT

My tent-mate that first year at Omo was Hank Wesselman, a fellow graduate student whose area of research was micromammals, those tiny creatures like bats, shrews, and rodents that are missed in even the most meticulous excavations of the kind that I was doing at Locality 398. Hank used a different technique. He took buckets of sediment from localities, soaked them in kerosene to break up the clumps, and then washed them through fine-meshed screens using a stream of water pumped up from the Omo.

I was interested in why Hank was studying micromammals when he was an anthropology graduate student. Rodents and bats are not particularly closely related to humans; they are not even primates. He explained that he was interested in the environments of the early hominids. Micromammals live in microhabitats, small areas of restricted ecological requirements to which they are adapted. Unlike larger mammals, they cannot range widely outside of their optimal habitat. The micromammal species discovered at a particular time and place in the fossil record are a very good indication of the range of habitats over which the larger animals, including hominids, roamed. Clever, I thought. This was where the new physical anthropology of Washburn was going under Howell's direction. The ecology of the fossil sites was being tested with increasing sophistication. The context of human evolution was being carefully investigated for the first time.

I began to look at my excavation with new eyes. I became interested in the possibilities that it held out for this type of paleoecological investigation. The small teeth of early hominids among the hundreds of other fossils became not so important as records of the anatomy of the hominids to which they had belonged but as place holders for their species in the natural scheme of things 2 million years ago. Hominids occurred at about the rate of one in a hundred identifiable mammal fossils. They were among the rarest of the animals.

A hundred new questions flooded into my mind. Why were hominids so rare? Had they been as rare in the environment as they were in the excavation? How had they died? How had their bones and teeth gotten into Locality 398, jumbled up with all the other animals? I had now ventured into the new field of *taphonomy*, the study of how fossils are formed and how they are buried and preserved.

Back in Berkeley I began to work on these questions in the lab. Along with Kay Behrensmeyer, a postdoctoral fellow in paleontology, I undertook a study in which we dropped an assemblage of human bones into an artificial stream, a "flume." Our goal was to determine how the various bones were sorted out by flowing water, much the same as the ancient Omo River must have done at Locality 398. We carefully weighed, measured, and recorded the behavior of each of the bones in varying speeds of water current. It became pretty obvious why there were no complete skulls at Locality 398. Regardless of the orientation in which they were dropped into the water, intact skulls held a water bubble in the front of their cranial cavity and they floated. The fast-flowing Omo River would have swept away any early hominid skulls from Locality 398 and deposited them in the calmer waters of Lake Rudolf, where Richard Leakey's team was waiting to discover them, I thought wryly. The densest parts of the human skeleton were the parts that stayed behind in the bed of the stream. These were teeth, exactly what we found in the greatest abundance at Locality 398.

At first the study of Locality 398 seemed to argue against the location's use in any serious reconstruction of paleoecology. Most paleontologists prefer to have a snapshot of a past time— a quiet burial place for fossils, which ideally are intact skeletons. Locality 398 was a more dynamic situation. Animals had died upstream; their carcasses had been picked clean by scavengers; their bones had been washed into the Omo by rain water; the river had swept them downstream; and the densest and most resistant parts of the skeletons had eventually been deposited in eddies of reduced river current on the inside bends of the river. Locality 398 was one such deposit. I believed, however, that this type of deposit was exactly the best kind to use in reconstructing the past environment. The processes that had formed the deposit had performed a statistically averaged, prehistoric census of the wildlife population in the general area, whereas a snapshot assemblage would have been only a view of one habitat.

My conclusion was that hominids had been rare animals in the environment, as rare as carnivores, such as leopards and lions. The implications of this finding were that hominids had been animals at the top of the food chain, species that had tended to prey on other species rather than being the prey species themselves. Their small numbers indicated that the survival of their offspring was much more assured than in species that reproduced in large numbers because of high mortality. In evolutionary terms, by this time hominids clearly had opted for quality of life rather than quantity of reproduction as a means of species survival.

The Locality 398 excavation also led to another important conclusion about the territory of early hominids. Ecologists have discovered that animal species that have low densities of populations range over bigger areas. These early hominids at

Omo, then, must have had large home ranges and probably covered up to several miles a day in foraging for food and traveling from one resting place to another. How territorial were they? Unfortunately, the answer to that specific question was beyond the scope of the excavation results, and other data could be interpreted in various ways. For example, the difference in size between males and females among early hominids suggested to some researchers that there had been a harem type of family organization, much the same as in savanna baboons, who defend a well-defined territory. But other researchers stressed the importance of the male–female pair bond and believed that early hominids may have been more similar to chimpanzees in having a poorly defined territory. The Omo results showed that the territory of early hominids was large, however it may have been defended.

## DEBATING THE AGE OF A NO-LONGER MISSING LINK

Following the 1973 field season at Omo, a conference was held at the Kenya National Museum in Nairobi. The purpose of the meeting was the presentation and discussion of the geology and faunas from around the Lake Rudolf basin. This area included the sites of Omo, east Rudolf, and a few older but smaller sites south of the lake. But there was also a hidden agenda at the meeting. A major storm was brewing over the dating of the contiguous sites of Omo and East Lake Rudolf. Some participants, my professor Clark Howell included, wanted to see a full-blown discussion of the issue. Others, particularly Richard Leakey, wanted to avoid this discussion and postpone it until a later time.

As a newcomer I did not notice that anything other than straightforward scholarly discussion was taking place until a rather extraordinary forensic ploy by Leakey gained my attention. Professor Basil Cooke of Canada had finished his presentation on the fossil pigs and Dr. Alan Gentry of the British Museum had finished his presentation on the antelopes when Leakey suddenly broke into the discussion, which was leading toward a consensus that he wanted to avoid. He picked up a glass pitcher of water and made an analogy with the problem at hand—something to the effect that the pitcher had real weight and substance but, yet, when viewed in the correct light there was no problem in seeing through it. Before the conference participants could change mental gears, Leakey suggested a short recess. And then, in order to subliminally undermine the effect of the last paper by the authoritative but soft-spoken Gentry, Leakey asked him to explain to the conference where the men's and ladies' bathrooms were in the museum. Taken aback, Gentry acquiesced. From my debating experience in secondary school I recognized these as transparent diversionary tactics, the first aimed at changing the subject and the second aimed at discrediting the source. What was going on here? The conference now had my undivided attention.

Cooke and Gentry had pointed out in methodical presentations that there was a discrepancy in the dating at Omo and east Rudolf. The fossils from east Turkana below a geological level known as the KBS Tuff did not match those from Omo that should have been of the same age. The KBS Tuff had been dated by two geologists at Cambridge University at 2.6 ± 0.26 million years ago using a new variant of the potassium-argon technique. Fossils from east Rudolf found in strata below the KBS Tuff, however, did not match those from Omo deriving from Member C, also dated to 2.6 million years. Instead the Rudolf fossils matched those from Omo Member F, dated at around 2 million years ago.

"Big deal," I thought. Revise the dates at Rudolf and see what went wrong with the new dating method. But it was a big deal to Richard Leakey. He had found the skull known as KNM ER (for Kenya National Museum East Rudolf) 1470 at a site below the KBS Tuff. Adding a few hundred thousand years for good measure, he claimed that he had found indisputable evidence for the earliest representative of the *Homo* lineage in the world, at nearly 3 million years ago. At this date 1470 was older, or as old as, most of the more primitive australopithecine hominids, which thus could not be ancestral to the *Homo* lineage. Thus the early date was very important to

Leakey in sustaining the pre-sapiens hypothesis inherited from his father. In fact, 1470 had been the focal point for a rapprochement between Richard and Louis Leakey, who had developed an increasingly competitive relationship, just before Louis died on his trip to the United States in 1972. Louis Leakey had hailed the early date for 1470 as a vindication of his longstanding belief in the antiquity of a uniquely human lineage separate from australopithecines. Richard was not about to give up easily.

## THE KBS TUFF DATING CONTROVERSY

After the Nairobi conference Richard Leakey moved swiftly. He relieved the scientists who did not agree with the early date for the KBS Tuff of their responsibility to study and describe the east Rudolf fossils. Basil Cooke was the first. Then Alan Gentry. Then Dick Hooijer of the University of Leiden, who had studied the fossil horses. John Harris, a paleontologist at the Kenya National Museum who studied the fossil giraffes, was put in charge of the pigs and antelopes. A visiting graduate student from Michigan named Tim White, eager to forge an alliance with the Leakeys, was to assist Harris on the pig study. A young French paleontologist was given the horse fossils to start on. Basil Cooke sent around a letter decrying such "scientific piracy," the expropriation of his data and observations that had already been publicly presented at the Nairobi conference. Those who were left on the east Rudolf project scrambled to develop hypotheses to make their data work with the older date.

One such hypothesis was that the ecology of the Omo fossil sites was significantly different from that at east Rudolf. Perhaps Omo harbored a relict fauna that held on in the perennial forest bordering the Omo River, whereas the lake margin habitats east of the lake were savannas and woodlands where only a more open-vegetation fauna survived. Thus the pigs, antelopes, and other animals at Omo would appear to resemble the fauna from an earlier period when there had been more forest cover at east Rudolf. This was the "lost world" hypothesis: the idea that a fauna could be ecologically insulated from change to such an extent that it would not evolve, sort of like dinosaurs still living in a valley in the Congo. This hypothesis would explain why the absolute dates at Omo and east Rudolf did not agree.

There were problems with the lost world hypothesis. First, all the fauna from Omo was far from forest-adapted. The large collections had turned up what was then the earliest record of camels in Africa, hardly a forest animal, and a significant proportion of the rest of the fauna—antelopes, pigs, horses, rhinos, birds, aardvarks—from all the various geological levels at Omo could be ascribed confidently to savannas or open vegetation habitats. Omo then preserved the same habitats as did east Rudolf. Second, evolutionary lineages of animals could be traced through the early levels at Omo to the later levels, using such measures as changing tooth size or progressive changes in the skull. If the east Rudolf fossils were plopped down into this sequence they would fit squarely at 2 million, not 3 million years ago. Third, the very small distance between Omo and east Rudolf, about a half-hour flight in a small plane, made the magnitude of the ecological difference postulated for the lost world hypothesis hard to conceptualize. The search for context that typifies modern paleoanthropology had spawned a debate, and, we thought, supported our position.

Our attention in Clark Howell's lab at Berkeley became focused increasingly on the single date of 2.6 million years ago for the KBS Tuff at east Rudolf. We thought the date was wrong. We spared no effort in trying to disprove it. We did statistical analyses of the entire faunas from Omo and east Rudolf and they bore out that the Rudolf fossils from below the KBS Tuff fit with Omo at 2 million, not 3 million years ago. Frank Brown reinvestigated his dating data from Omo. There were potassium-argon dates stacked on top of potassium-argon dates all through the sequence. The stratigraphic framework had been checked and rechecked by Belgian geologist Jean de Heinzelin. A further test on the sequence of reversals in the earth's magnetic field, paleomagnetism, was undertaken in the field to confirm the absolute dates from potassium-argon.

Everything fit. There was definitely something wrong with the east Rudolf story.

The Watergate scandal was just winding down and Richard Leakey was adopting a Nixonesque stonewalling approach to his own problems at east Rudolf. He ordered all scientific reports by his team to be held in house, subject to his review. A doctoral thesis completed at the University of London on the geology at east Rudolf was declared off limits to our team at Berkeley by Leakey. We got around him; we simply had a friend go to the university library in London, photocopy the thesis, and send it to us. Leakey refused all requests for another lab to undertake an independent potassium-argon analysis of the KBS Tuff. A geology graduate student at Berkeley finally managed to get some samples of the KBS Tuff, through the intervention of archaeologist Glynn Isaac, Leakey's colleague at Berkeley, and brought them back to Berkeley to Garniss Curtis's lab for analysis.

Everyone waited with bated breath for the results from Curtis's lab. When they came in they answered some questions but raised others. The main KBS Tuff was indeed 1.8 million years old, the age that our team had been convinced of. But samples from the so-called KBS Tuff in another area gave a different age. This was not the KBS Tuff at all, but another layer that resembled it. The two had been mistakenly correlated by the field geologists at east Rudolf. If marker horizons like the KBS Tuff had been misidentified, then there were clear stratigraphic problems at east Rudolf to add to the dating problems. The fossils from east Rudolf were of little help in sorting out this confusion because they had been collected not from point localities recorded on aerial photographs, as at Omo, but only within vast areas that cut across several large geological time intervals.

Leakey and his team minimized the impact of the redating of the KBS Tuff. His support for an ancient, separate lineage of *Homo* gone, Leakey nevertheless stood by his version of the pre-sapiens hypothesis. But at the same time he quietly encouraged all the members of the geological team to find research projects elsewhere. Even Kay Behrensmeyer, whose initials had given the KBS Tuff its name, left east Lake Rudolf, now renamed Lake Turkana, for research in Pakistan. Leakey then adopted the policy of "if you can't beat them, join them." He asked Frank Brown to do the geological work—the dating and correlation between areas—east of the lake.

Frank Brown has an attachment to the land, in this case the East African Rift Valley, like few men in the modern world know terrain. Our ancient hunting ancestors knew every rock, rivulet, glade, plant, and animal of their territory. Today we tend to be more like trained laboratory rats, running sterile mazes of highways and sidewalks. When Frank and I first arrived at Omo he gave me a first-gear, four-wheel-drive tour of the Omo—through dry stream beds, across trackless bush country, into ravines, up cliffs, and back to camp again. Just like the Australian aborigines who have names for all the geographical features in their territories, Frank had names for all the rocks—Tuff A, Tuff B, Tuff C, and so forth. He had compiled a book of all the plants found in the Omo. And he would suddenly stop to point out a bird. "Variable sunbird" or "carmine bee-eater," he would say. I realized that he liked driving over all the bumps, into all the depressions, around all the obstacles, like a man experiencing the texture and curves of a lover's body.

When Frank Brown called me years later on an early Sunday morning in New York to ask my opinion of whether he should accept Leakey's offer to work on the east Turkana geology, I immediately said that I thought he should do it. He was surprised. Of all the Berkeley group I had been the most critical of Leakey and his handling of the research agenda at East Turkana. (I learned later that I had been given the appellation the "mad dog of Berkeley" for my persistence in pursuing research questions that Richard Leakey and his team would rather not have had examined in the detail that we did.) But Leakey's team had succeeded in discovering extremely important fossils, which I had studied in Nairobi in great detail, and they deserved to be put into a clearly understood context. Frank could do that better than anyone else I knew. I also knew that he had an unbreakable covenant with the land.

## A MAJOR DISCOVERY AT OMO

The most important and complete early hominid discovery at
Omo was made by one of Howell's surveyors, Januari
Kithumbi. One morning late in the 1973 season Kithumbi
found some flat bits of bone at a locality far in the north known
as Locality 894. In the afternoon, Howell and I went back there
with a group to look for more pieces, because the fragments
that Kithumbi had found were parts of a hominid skull. There
were a few more pieces on the surface, and we found some
more when we screened all the sediment around the find. All
the pieces were wrapped carefully in toilet paper and went with
Clark directly back to Berkeley. The next year when I was in
the Omo without Howell I excavated the entire deposit at
Locality 894 but we found only a few more fragments.

Putting together the hundred or so fragments of bone and
teeth of L894-1, as the skull was named, was a task that took
me well over a year. The breaks were sharp and when I found
fits they were firm. But the fragments came from all parts of
the skull, indicating that the specimen had been complete when
it lay preserved in the ground. I thought of the Locality 345
baboon skull. It had been similarly complete, and we had excavated that specimen before erosion had touched it. We had discovered Locality 894 too late to prevent the rains from
exposing the specimen, breaking it up, and washing many of
the fragments away. Of course if erosion had not exposed the
bones in the first place, Kithumbi would never have found the
site. Many of the pieces had been transported too far away to
ever find again, and large parts of the skull appeared to be
missing.

My job was to put together as much of the skull as I could
and figure out what hominid species it belonged to. The teeth
were present, unlike most of the well-preserved skulls from
east Turkana, and this helped immensely in diagnosing what
early hominid we had discovered. As in other vertebrate fossil
groups, teeth are important in typifying and identifying a
species. The molar and premolar teeth were in a human-like
harmonious relationship to one another, like the gracile

*Australopithecus africanus* and very unlike the hugely inflated back teeth of the robust *Australopithecus boisei*. Other characteristics of the skull and face helped us to identify the L894 hominid. The skull vault looked small, globular, and rounded, unlike the low, flat, keeled skull of *Homo erectus*. And the frontal bone showed no projecting brow ridge so characteristic of *Homo erectus*. L894-1 seemed to be in the middle between *Australopithecus africanus* and *Homo erectus* and to share many of the characteristics that Louis Leakey, Phillip Tobias, and John Napier had used to define the new species *Homo habilis* from Olduvai Gorge. In 1975 Howell and I ascribed the specimen to *Homo habilis*. It was the first fossil hominid to have been ascribed to *Homo habilis* found outside Olduvai Gorge.

The L894-1 skull came from a geological level high in the Omo sequence, upper Member G, dated to just under 2 million years ago. At that time Lake Turkana had expanded to the north, and this hominid had died on the shores of the lake. Invertebrate fossils known as ostracods found associated with the site indicated that the lake was salty at that time. This meant that the habitat had been open savanna, with few trees. Like Lake Turkana today, the fluctuating saline lake waters would have killed trees growing along the lake margins. We had proof that *Homo habilis* had been a savanna-living hominid.

L894-1 was exactly the same age as Richard Leakey's well-known fossil hominid skull known as ER 1470, and it had lived around the margins of the same lake. After studying all the east Turkana fossil hominids, I was convinced that 1470 and several other similar fossils were, with L894-1, members of the same population of *Homo habilis*. They had lived cheek-by-jowl with the robustly built *Australopithecus boisei*, the only time in hominid evolution when two contemporaneous species of hominid coexisted. The robust australopithecines died out about a million years ago. The *Homo* line persisted to eventually evolve into modern humans.

It had taken me quite some time and a lot of study to arrive at these conclusions. One lesson that I had learned from struggling with the Omo and Lake Turkana hominids was that it is devilishly difficult to figure out what a fragmentary specimen really is. And sometimes the specimens weren't so fragmentary. Take the east Turkana specimen ER 1805, for example. This was a small, rounded skull that looked like *Homo habilis*, but it had a crest of bone that ran front-to-back along the top of its head, known as a sagittal crest. Previously known *Homo* did not have this trait. Did 1805 represent a new species of *Homo* with a sagittal crest? I didn't think so. I thought back to the lessons of the new physical anthropology and remembered that the variability of individuals within populations is important to consider. Was there a functional reason that 1805 might have a sagittal crest? I noticed that the specimen also had an abnormally wide palate, despite its small teeth, and I suspected that its chewing muscles had been large. My interpretation was that 1805 had been a relatively small-headed *Homo habilis* with excessively well-developed chewing muscles that had met in the midline at the top of the skull to form a sagittal crest. Pre-Cold Spring Harbor interpretation of 1805 would have been that it represented a separate species, if not genus, of early hominid.

Once my study of the Omo and east Turkana hominids had been completed and my doctoral dissertation had been turned in, I still had many questions. Where and when had these lineages originated? They certainly filled in some of the gaps between the earliest hominids and human beings, but they were not the missing links leading back to the common ancestor that we shared with the apes. The answers lay in earlier rocks, in more primitive hominids, and possibly in different areas than where we were now looking. I started by restudying in detail the earliest hominids known, the early australopithecines from East and South Africa.

# 4

# Problems with Dating an Older Woman

It was late fall 1974 and I was sitting in my room at the Ethiopia Hotel in Addis Ababa when the phone rang. I was mildly surprised because I did not expect any calls. The negotiations that I was involved in with the Ethiopian Ministry of Culture were all undertaken face-to-face and I was not even aware that anyone there knew where I was staying. In any event, I had only left the ministry's offices a few hours before. "Probably the front desk about a problem with my laundry," I thought. I picked up the phone. It was Don Johanson and he was excited.

Don Johanson and I had never been close friends. He had gotten off to a rocky start with me when Clark Howell had introduced us the year before at the national physical anthropology meeting in Dallas. Clark had said by way of introduction that I would be coming to the Omo that year, to which Johanson had condescendingly replied, "we'll keep you busy." But as things worked out Johanson had been kept at the University of Chicago during that Omo field season by dissertation problems, and he and I had had only a few casual meetings since, which had been cordial. He was supposed to be in the field at his new site in the north of the country at Hadar, and I was surprised to have a call from him.

I asked him what was going on. The country was tense because Emperor Hailie Selassie, the original "Ras Tafari" on which the Rastafarian religion was based, the organizer of the Organization of African Unity, and the all-powerful ruler of Ethiopia, was under house arrest. The rumor was out that Ethiopia was ready for revolution. But Johanson said that his call concerned none of that. They hadn't had any problems. Instead he asked me if I remembered Alemayu Asfaw, a representative from the Ethiopian Ministry of Culture whom I had begun training in field surveying in the Omo earlier that year. Alemayu had been a youth and the Omo had been his first assignment. I said yes, I remembered him. Johanson then said that a few days before Alemayu had found an incredible cache of hominids at Hadar. He had just reached Addis with the fossils in order to prepare the documents for study export and to get a press release written up with the ministry. Would I like to see the fossils? Johanson clearly wanted to show off the specimens, and I certainly wanted to see them, so I said "You bet."

## A QUESTION OF IDENTITY

Don Johanson arrived in about half an hour, and we sat down in a corner of the lobby of the Ethiopia Hotel to look at his new discoveries. He pulled them out of several cigar boxes and carefully, reverently, unwrapped them. There were two jaws and two halves to a palate that fit together. The preservation of the bone and the teeth was magnificent. I remarked on this. The Omo and Richard Leakey's site of east Rudolf in Kenya both tended to have fossils with bone extensively cracked and weathered from months or years in the sun and rain before they had been covered by sediment and fossilized. The Hadar bones must have been buried quite soon after death to be so well preserved.

But what were they? To what early hominid species had they

belonged? These were the burning questions that Johanson wanted to answer. I suspected that he had come to bounce his ideas off me, probably a run-through for the press conference or his planned meeting with Richard and Mary Leakey in a few days.

Johanson pulled out several sheets of graph paper on which he had plotted the dimensions of the teeth of the new specimens from Hadar. He had used a red pencil to outline the front teeth as a group and then the back teeth as a group. The new specimens matched the size ratio of front to back teeth seen in *Homo sapiens*, our own species. The Hadar teeth were clearly different from the large-jawed parahumans, the robust australopithecines, which had huge molar and premolar teeth and relatively tiny incisor and canine teeth in the front of their mouths. I could see where he was going. Johanson wanted to make these specimens out to be *Homo*.

I didn't want to throw cold water on Johanson, basking in the warm glow of his newly found hominid fossil treasures, but I had to point out that the relationship of front tooth size to back tooth size would not serve to determine whether or not the Hadar specimens were *Homo*. While this ratio was very useful for sorting out the robust australopithecines, it did not sort out *Homo* from another type of early hominid, the lightly built, or gracile, australopithecines. With all the isolated early hominid teeth I had begun to deal with from Omo, I had read and reread John Robinson's classic monograph on the South African fossil hominids, *The Dentition of the Australopithecinae*. If one conclusion came through loud and clear, it was that gracile australopithecines had a harmonious dentition—front teeth in the same size relationship to back teeth as in modern humans—and robust australopithecines had a disharmonious dentition.

He persisted. "Look at how similar the palate is to Pith 4," he said. He was referring to Pithecanthropus IV, a specimen classified as *Homo erectus* from Indonesia that I had recently studied at its museum in Germany. "Yes," I said, "but Pith 4 has some uncharacteristically primitive characteristics." Among them was a gap in the tooth row, a diastema, in front of the canine tooth.

Much had been made of this ape-like characteristic in this specimen, and now the Hadar palate had the same character. In apes the diastema in the upper jaw is the space into which the large lower canine tooth from the lower jaw fits when the jaws are closed. By that time, most anthropologists had accepted that the diastema of the Indonesian specimen was a fluke of variation, sort of like a human baby being born with a tail—certainly a primitive trait but not characteristic of the species. I thought that the diastema in the Hadar specimen, on the other hand, might be a truly primitive characteristic. But I soon realized that Johanson had already made up his mind. He was going to call the Hadar specimens *Homo*.

After Johanson left, I puzzled over his insistence. Why would someone who had studied the South African gracile australopithecines, as I knew Johanson had, want to pass over the obvious similarities and ascribe the Hadar specimens to *Homo*? I concluded that the answer must lay with the Leakeys—Louis, Mary, and Richard—who had remained committed for years to the idea of an ancient *Homo* lineage running alongside the more primitive *Australopithecus*—a sort of African pre-sapiens hypothesis. Don Johanson now had "fossils dazzling enough to match those of paleoanthropology's certified supernova, Richard Leakey," as he wrote in his book *Lucy* in 1981. But he also felt he had an edge, something more: fossils that were much older than Leakey's. Comparisons with fossils from the well-dated levels at Omo showed that the Hadar hominids were about 3 million years old.

When Richard and Mary arrived shortly thereafter, Johanson took them on a tour of the Hadar deposits and showed them the new fossils. "All in all, I'd have to call them *Homo*," Richard is reported to have said to Don. "So would I," said Mary. Expectation fulfilled, Johanson rushed to the conclusion. "If that diagnosis stood up, it meant that these were the oldest human fossils in the world" (*Lucy*, p. 176).

Taxonomy and chronology—naming and dating—were the two critical issues. Louis Leakey's missionary father might have intoned, "on these two commandments hang all the law and the prophets."

Richard Leakey had his own reasons for wanting to see Johanson's fossils categorized as *Homo*. There were dating problems at Leakey's site east of Lake Rudolf, now renamed Lake Turkana by the Kenyan government. Leakey's prize skull, known by its museum accession number of 1470, had been collected below a geological level known as the KBS Tuff. A tuff is a deposit formed of ash erupted from a volcano and typically has minerals within it rich in the element potassium, just like potash from a campfire. Potassium is important because the rate of its clock-like decay into argon forms the basis for dating how old the rock layer is.

Leakey had proclaimed the discovery of the 1470 skull several years before as the earliest example of the genus *Homo*. Most anthropologists had agreed, and Leakey had been lionized by the press. I remembered the New York meeting of the American Anthropological Association in 1971 when Richard Leakey had made his debut with 1470. The atmosphere had been electric. Old paleoanthropologists in black suits talked excitedly in raspy voices, and all of them had opinions on where to classify 1470, most squarely in the genus *Homo*. But where to classify Richard Leakey was a different problem. Loosely associated to the Kenya National Museum, without university degrees, and with no formal university ties, Richard Leakey had been an outsider here, commended by a paleoanthropological establishment as a wonderfully successful "collector." But the press had no such qualms and promptly labeled him as a paleoanthropologist. Johanson, I concluded, was clearly of the latter opinion as well.

## THE GEOLOGICAL AGE OF SKULL 1470 AND THE HOMO LINEAGE

When skull 1470 was found it was initially thought to date to around 2 million years ago, or 200,000 to 300,000 years earlier than the oldest, Bed I, levels at Olduvai Gorge. This estimate was based on the evolutionary stages of the animal fossils found associated with the hominid skull in the area at east Turkana in which it had been found. Leakey's team collected fossils in vast areas, numbered consecutively. 1470 had come from Area 131 and the fossil animals found there were comparable to the faunas found at other African sites, particularly Omo, dating to about 2 million years ago. But later Leakey's team revised the dating of 1470—incorrectly, as discussed in the last chapter—to a date of nearly 3 million years old. This date was astoundingly early for the appearance of the genus *Homo*. They based the revised dating on an absolute date of the KBS Tuff using a new version of the potassium-argon technique, argon 40-39.

Dating hominid discoveries had become an important issue since the advent of the new biological anthropology. In the old days, the anatomy of the fossil and its general geological age sufficed to tell anthropologists everything they wanted to know. But now the scenarios of evolution were too complex. The anatomy of the fossils was dissected and statistically analyzed in such excruciating detail and the context of the hominids was collected and plotted in space so methodically that the whole nature of the science had changed. Great care is now needed in fieldwork to record accurately all the fauna from a site, not just the hominids. Unfortunately at east Turkana this recording was not done initially. Little concrete markers were erected to accurately mark the site of discovery of hominids, but no such attention was paid to the animal fossils found within areas. A whole rhino skull, rarer even than a hominid skull, might be labeled only "Area 104." Because a number of different strata were exposed in Area 104 this designation would not be sufficient for scientists at the museum to figure out how old the fossil was.

The uncertainty in the exact location of fossils at east Turkana caused problems. When he saw the records of the collections housed in the Rudolf Room of the Kenya National Museum in Nairobi, one senior paleontologist who had been enlisted as a member of Leakey's team called the project "a paleontological disaster." Leakey asked a young and energetic vertebrate paleontologist from Princeton, Vincent Maglio, to come in and attempt to rectify the situation.

Maglio began by making new collections and accurately recording the areas and the relationships of the fossil finds to particular geological marker beds within the area. For example, the records for specimens now would read "Area 131, below the KBS Tuff." After some time Maglio was able to build a framework of four faunal zones at east Turkana, labeled I through IV. These zones were named for one of the characteristic species found within that zone. For example, the pig species *Mesochoerus* [now *Kolpochoerus*] *limnetes* lent its name to Zone III. Maglio was able to put the zones in an order of oldest to youngest by following the evolutionary stages of the animals in the zone. Zone III, for example, had bovids (antelopes and their relatives) that were more evolutionarily advanced than in Zone II. Therefore, Zone II was older than Zone III. The fossils that were first collected without geological information were then plugged back into this system of faunal zones in order to gain some idea of where they fit in the fossil sequence at Lake Turkana.

Although this exercise in salvage paleontology horrified scientists used to the exacting standards of collection management at the Omo, the fossils of which were housed in the same building on the national museum grounds in Nairobi, in the end this was not the primary problem with dating the east Turkana fossil levels. The real problem came with the geology. Here the contrast with Omo was even starker.

Much of the difference in the geological interpretation of Omo and east Turkana can be traced to the nature of the exposed rocks in the two areas. Omo is an area of badlands reminiscent of southern Utah, with high cliffs, bluffs, and buttes—high exposures that reveal the strata in sequence. Each exposure is slightly different, and Jean de Heinzelin had played a masterful game of matching and overlapping the sections to produce a general stratigraphy of the Omo beds. Every site could be positioned exactly on the master stratigraphic chart. But east Turkana had a different topography. It was an unimpressive area of low, nondescript hills and washes. There were no high hills with extensive exposures. Geologists at east Turkana tried the game of stratigraphic match and overlap over and over, but they failed to produce a consistent framework. Then Richard Leakey shuffled the team and Frank Brown took over the geological research program at east Turkana.

Brown quickly arrived at a plan of action. The lack of good exposures and the poorly controlled paleontological collections at east Turkana meant that an unambiguous geological framework had to be established before some sense could be made of the whole. He decided that the way to characterize the various tuffs at east Turkana was not by way of the very limited evidence presented by their external appearance but by their chemistry. He proposed to identify each tuff by the unique mix of chemicals that characterizes an eruption from a volcano. Evidence from volcanoes around the world indicated that tuffs derived from different volcanoes differed markedly in chemical composition, as did tuffs erupted at different times from the same volcano. Brown could obtain a chemical fingerprint of each tuff and thereby get a good handle on what each tuff was at east Turkana. His results fully confirmed the original faunal dating of 1470 at about 2 million years old, and the potassium-argon redating of the KBS Tuff as 1.8 million years old.

Leakey garnered kudos from the professional paleoanthropological community for bringing Brown on board to oversee the geological research program. But none of the findings Brown made in terms of the age relationships of the hominid fossils made it into Leakey's paleoanthropological pronouncements. He never changed his position on an ancient lineage for *Homo* and he never varied the text of any of his popular treatments of hominid evolution. He just stopped talking about any specifics relating to dates or lineages of hominids.

This evasiveness became very frustrating for Don Johanson, who needed Leakey for a target. In a three-way interview undertaken by Walter Cronkite with Johanson and Leakey, Leakey had drawn a large X through his side of a proposed diagram of hominid evolution. He wasn't prepared to talk about specifics. He didn't think the data warranted being specific. Johanson was left jousting with a phantom.

## THE HOMINIDS FROM HADAR

But other people were talking about the specifics of the hominid fossil record and were drawing conclusions about human evolution. After having studied in detail the hominids from Olduvai and east Turkana at the Kenya National Museum, I, for one, wanted to know the specifics about Johanson's Hadar hominids. I went to the Cleveland Museum of Natural History, where Johanson was based, to study in a more systematic manner what Johanson had first shown me in the hotel lobby in Ethiopia. Soon after our meeting in Addis Ababa, Johanson had returned to the field and had discovered "Lucy," a partial skeleton of one of the hominids that had died at Hadar. I particularly wanted to see this specimen.

Johanson gave me a hospitable welcome. He was grappling with the Hadar hominids. While he had been ecstatic at the discovery of Lucy, she had given him some real difficulties. Lucy was the spitting image of *Australopithecus africanus*, the South African gracile australopithecine dating to around 2.5 to 3 million years ago first discovered by Raymond Dart and Robert Broom. She was somewhat smaller in size, but her mandible and pelvis were very close anatomical matches. And the Hadar hominids were considered in the same age range as *A. africanus*, 2.8 to 3.2 million years old. Johanson had trouble putting Lucy in the same category of *Homo* to which he had decided his palate and mandibles belonged. And then he showed me a piece of a humerus that was much larger than Lucy. Johanson thought that this specimen might indicate the presence of a robust australopithecine at Hadar.

After several days of examining, measuring, and comparing the Hadar hominids, I had an idea of what they were. The impression that I had had in Addis became more firmly established in my mind. These hominids were East African versions of the South African *Australopithecus africanus*. Were all the specimens from the same population? I thought so. First, the range of variation in body size was large but not outside what was observed in other large primates, such as gorillas and savanna baboons. Second, the pattern of variation between specimens did not seem to indicate a different pattern of variation that could characterize different species. In other words, the individuals at Hadar seemed to show the variation that one might see in any population that had members of different ages and different sexes, but not the variation that one might see between individuals of different species. And finally, the geological conditions under which the Hadar hominids had been deposited suggested to me that they had all lived, died, and been buried in one constrained area and in one restricted time period. In the absence of strong anatomical indications of different populations, this was further support that we were dealing with one population.

Johanson listened to my ruminations about his specimens. We had some good discussions over the comparisons with other hominids. He seemed most interested in my thoughts on the Hadar hominids belonging to one population. Walking out of a Cleveland restaurant after lunch he remarked on what a big job the description of the Hadar hominids was. I realized that this might be an opening for me to offer to help. I held back. Maybe it was his remark in Dallas or maybe it was my feeling that Johanson wanted more agreement from me than he was likely to get. In any event, I steered away from the topic.

Johanson published the Hadar hominid discoveries with interpretations that ran the gamut. In 1977 he published three short papers that purported three different interpretations of the specimens. One said that they were all *Homo*. One said that they were a mixture of *Homo*, *A. africanus*, and *A. robustus*. And one said that they were all *A. africanus*. But by the following year the interpretation had solidified. Johanson had forged an alliance with Tim White, a recent Ph.D. from the University of Michigan who had been working on the early hominids from Mary Leakey's site at Laetoli, Tanzania.

Johanson needed an interpretation that could somehow amalgamate his various observations. He and White decided that at Hadar they were dealing with one population, but a population of a species new to science. Furthermore, they concluded that this new species was also present at Laetoli. They named the new species *Australopithecus afarensis*, taking the species name from a locality name at Hadar, but they chose as the type specimen a mandible from Laetoli. Mary Leakey had been a coauthor on the paper naming the new species, but when she saw proofs that named it not a species of *Homo* but of *Australopithecus*, she withdrew her name.

Johanson and White emphasized the primitive aspects of the Hadar specimens. Perhaps the most controversial part of their argument had to do with *Australopithecus africanus*. I had published my observations, along with independent observations by South African paleoanthropologist Phillip Tobias, that compared Hadar and Laetoli hominids closely with *A. africanus*. Johanson and White emphasized the differences and suggested that the Hadar specimens had little to do with them. In their interpretation of the lineage of the new species *afarensis*, they hypothesized that it was uniquely ancestral to *Homo*. This conclusion, of course, tended to vindicate Johanson's earlier claims for *Homo* status. Johanson and White noted a large degree of variation due to male–female differences in the Hadar hominids, a type of variation known as sexual dimorphism. This explained Johanson's earlier claim that the humerus and maybe a temporal bone as well had indicated the presence of robust australopithecines at Hadar. They now interpreted these as males of *A. afarensis*. Finally, they thought that *A. africanus* was uniquely ancestral to *A. robustus* and had nothing to do with the ancestry of *Homo*. I strongly disagreed with this latter conclusion most of all. I thought then that their naming a new species and their phylogenetic hypothesis had more to do with saving face than with propounding the most conservative scientific interpretation.

The fossil sites of Hadar, Ethiopia, and Laetoli, Tanzania, are over one thousand kilometers apart. Johanson and White had used a specimen from Laetoli and tacked the Ethiopian site name on it. It is frequent practice in zoological taxonomy to apply a locality name to a species or subspecies. This name usually reflects exactly where the type specimen was found and no interpretation is involved. For example, if a new earthworm species were to be discovered in the Budongo Forest, it could be named *budongoensis*. If a fossil were found representing an extinct but previously unknown species at Olduvai Gorge, the species name could be *olduvaiensis*. But the mandible from Laetoli had not been found at Afar, and thus Johanson and White's interpretation, that Hadar and Laetoli hominid samples were the same species, was required. Why had they done this?

I surmised that Johanson and White were not confident that

the Hadar sample truly represented a new species. They thought that Tobias and I might be right when we ascribed it to *Australopithecus africanus*. The much smaller Laetoli sample, composed primarily of teeth and jaws, however, was up to a million years older than Hadar and the *A. africanus* sites in South Africa. If the type specimen were to come from this sample it had a much higher likelihood of being from a more primitive population and hence different from *A. africanus*. *Afarensis* would go down in history as a valid species name even if Lucy and her kin were to be put into *africanus*. And by putting both samples into *afarensis* Johanson and White had succeeded in creating their own terrain for the debate. The Hadar sample was interpreted more in light of the primitive Laetoli specimens than the more advanced *africanus* specimens.

## DATING OF AFARENSIS

The relative dates of Hadar and Laetoli became important for the same reasons. If Hadar were as old as Laetoli, then both sites were older than the South African sites from which *A. africanus* was known. While geological age per se was not a diagnostic character in taxonomic naming of species, similar age was certainly a bolstering prop to the same-species argument for the combined Hadar-Laetoli fossils.

In the mid-1970s, geologist Jim Aronson, a rumpled, friendly, and incorrigibly informal professor of geology at Case Western Reserve University, wandered into the supercharged atmosphere of early hominid pyrotechnics. One can imagine Don Johanson driving across town in Cleveland and asking him nonchalantly if he'd like to go to the Ethiopian badlands, collect some rocks, and run them in his lab. Aronson did. And his results, reported in 1981 in the British journal *Nature*, set off some fireworks of their own.

Aronson and his colleagues dated a basalt, an ancient lava flow, at Hadar. The basalt was called KMB, for Kadada Moumou Basalt, a name that I had always thought tripped easily off the tongue. But Aronson's date left me speechless. The number that came out of his lab for KMB was 3.6 million years ago. Was this going to be a replay of the KBS Tuff controversy? If the Hadar hominids were this old, they were the same age as Laetoli. I had a hard time accepting the date because the faunal similarities between Hadar and Omo pointed unambiguously to 3.0 million years ago.

Ever since Cold Spring Harbor the possibility had existed that the various disciplines in the new multidisciplinary biological anthropology might not agree with one another. What would happen? Which one would win out? We were in one such contretemps now. Paleontology gave one date and geochemistry gave another. To resolve the issue, the scientists involved need to delve into the assumptions and limitations of the other discipline to put their findings into perspective. This process makes for debate, sometimes heated and sometimes perceived as a little naive from the perspective of the other discipline, but the science of human evolution benefits from this cross-fertilization.

I was interested in the accuracy of the dating of Hadar for much more than principles of proper accountancy in the ledger book of hominid paleontology. At the time of Aronson's dating of KMB I had just finished a long paper on the status of Stephen Jay Gould's idea of "punctuated equilibrium" in the human fossil record. Punctuated equilibrium is the thesis that evolutionary lineages went through long periods of little or no change and then underwent rapid bursts of evolutionary change to become new species. In Gould's lingo this was "stasis" interrupted by "punctuational events." Early on he had touted this newly discovered mode of evolutionary change as an alternative to Darwinian evolution, which he labeled gradualism. Darwin believed, and it is still generally accepted, that evolution proceeds in small steps, making evolutionary change a gradual trend.

I had thought since graduate school, when Gould had come to Berkeley for a talk, that his conclusions were far out in front of his data. I put him into that category of scientist who is more scholar than empiricist. Was it any coincidence that even the name of the phenomenon that he claimed to have discovered was an analogy to the written word? "Punctuation"—it was as if Stephen Jay Gould could write it and make it so.

Gould had claimed that the hominid fossil record was one of the best examples of the phenomenon of punctuated equilibrium that was available in the fossil record anywhere. When my friend Jack Cronin, a molecular bioanthropologist then at Harvard but trained at Berkeley, suggested that we test Gould's notion, I thought it would be an interesting and important topic to investigate.

The question of punctuated equilibrium in evolution is really just a question of rate of evolutionary change, and the data that you need to test this are accurate dates for the fossils. In our analysis we used the correct dates for the east Turkana hominids. Gould had used the nearly 3-million-year-old date for Richard Leakey's *Homo habilis* skull, 1470, and claimed that no change had occurred for over a million years, or until after Olduvai Bed I, dated at 1.8 million years ago. The redating of 1470 to 2.0 million years ago removed this part of his argument. 1470 and Olduvai Bed I *Homo habilis* were clearly close in morphology and they were also not distant in terms of time. We further made the argument that 1470 was slightly more primitive in some of the characteristics of the face and skull than later *Homo habilis*, an observation that fit well with a Darwinian model of gradual evolutionary change because 1470 was slightly earlier in time.

The Hadar early hominid fossils had come into the picture of punctuated equilibrium because Gould, using the original dates for both sites, had used them to postulate that there had been only slow evolutionary change—stasis—in *Australopithecus afarensis* from Laetoli to Hadar, 3.7 to 3.1 million years ago, using average ages of the two sites. Because of the fragmentary nature of the Laetoli hominids, one of the measures that we were testing over time, brain size, could not be obtained. No skulls in the Laetoli collection could give us a cranial capacity. So we had used two cranial capacities from Hadar specimens and plotted them at 3.0 million years on our graph of hominid brain size growth over time. Plotting the few dozen specimens with good cranial capacity values on a graph we found a clear, steady correlation between age and brain size. Although brain growth speeded up in the genus *Homo*, over the whole length of hominid evolution, for every 500,000 years of hominid existence, brain size grew by 250 cc. Such steady growth supports Darwin's gradual mode of change and contradicts Gould's model of punctuation. But if the date of Hadar were suddenly to be revised 600,000 years downward, the results of our test would significantly alter. This is why Aronson's new dates on KMB arrested my attention.

Fortunately, geologist Frank Brown became interested in the problem. Some time before, I had visited Frank at the University of Utah and had given his introductory class in historical geology a brief overview of some of the major issues in hominid evolution. I talked about geologically appropriate topics like dating and fossil collecting, and I touched on the theoretical issue of punctuated equilibrium. I thought I was being interesting, but the lecture fell flat. Seldom have I had to talk over snoring. So I finished early, apologizing for not keeping the students the entire hour. But at least Frank had listened to the talk. Whether or not he gathered from this lecture that I would be interested in new dating results from Hadar, I don't know. But I was very interested when he called me to say that he and some colleagues had used his tuff fingerprinting method to pin down dates for Hadar.

Frank Brown had contacted Aronson at Cleveland to get some tuff samples on which to run chemical analyses. Aronson had been very obliging. Unlike the problems with obtaining samples for dating the KBS Tuff, Hadar was not going to have to be another "Turkana-gate," I thought. Working with a group at the U.S. Geological Survey in Menlo Park, California, Brown found that there were some trace element chemical matches between the tuffs at Hadar and those in the Turkana basin. The presence and amount of each trace element was distinctive for each tuff, which is true even of tuffs erupted at different times from the same volcano. Brown and his colleagues interpreted the results to mean that the same geological levels, which were well dated in the Omo and east Turkana, were also present at Hadar. This finding meant that a volcano somewhere in central Ethiopia had exploded and sent ash in both directions, depositing a tuff in both the Turkana and Hadar regions. It also meant that an independent test of Jim Aronson's dates was now possible.

When Frank Brown called to tell me the preliminary results of his tuff analyses, I was not surprised. The initial dates as indicated from the fossils were right. The tuff overlying the Lucy site was the same tuff that in the Turkana basin had been dated at 2.8 million years old. Lucy was therefore close to 3 million years old, and nowhere near 4 million years old, as Don Johanson was claiming on the basis of Aronson's revised dates.

"What do you think their response will be?" Frank asked me, referring to Don Johanson and Tim White. "I think they will fight it," I said. "But how can they?" he replied. "I don't know," I said. One possibility was a variant of the old lost world argument used at east Turkana—that Hadar and Turkana had the same faunas, but ecological similarities between the two sites had masked the fact that they were truly different in age. Frank was incredulous. "Okay," I continued, "how rock solid are the individual identities of the tuffs, even from the same volcano? Could you ever get confusion between two tuffs erupted at different times from the same volcano?" He said, "very unlikely," but he agreed that it would be useful to get some quantitative backup ready.

Then Frank called Tim White and sent him and Don Johanson the preliminary results. White lost no time in responding. The correlations had to be wrong. They did not believe that tuffs could be spread so far as between Turkana and Hadar. Even if they had been, they did not believe Brown's technique could effectively distinguish different tuffs from the same volcano, which could erupt many times over several millions of years. White also said—surprising me because it was a new claim—that the Hadar fauna correlated more closely with Laetoli than with the Turkana basin. "What do you think of that?" Frank asked me. I said that I thought the assertion was flatly wrong. I said that I would pull together the Turkana basin fauna from between 2 and 4 million years ago and do a detailed comparison just to see if my impression was correct.

To counter the argument that the tuffs at Hadar and Turkana were only local phenomena and not widespread, Brown and colleagues looked in the deep sea core off the East African coast for potential matches. They found one in the Gulf of Aden. It was dated to less than about 3.3 million years ago, in accord with Brown's results. The wide extent of the tuff was only surprising to anthropologists who had been used to thinking of Hadar and the Turkana basin as totally separate entities. In fact, ash from an erupting volcano can be spread hundreds and even thousands of miles by the wind before being deposited.

For my analysis I began with the Omo computerized catalog, which recorded every fossil from every locality in the Omo valley of the Turkana basin. I contacted Clark Howell at Berkeley, who had directed the research at Omo, and he agreed that the project would be important. We used a number of statistical formulas to measure the faunal similarities between Omo and Hadar. Then we looked at individual fossil species one by one to determine at which other sites they were found. Our results showed that Hadar was closest to Upper Member B at Omo, dated to 3.2 to 3.3 million years ago. The analysis confirmed the dating by Brown.

The papers reporting the tuff correlations between the Turkana basin and Hadar by Frank Brown and colleagues, and the faunal comparison that I had undertaken with Clark Howell and one of my graduate students, Monte McCrossin, were published together in the British journal *Nature*. There was substantial interest in the revised dates for Lucy. British paleoanthropologist Michael Day wrote an editorial in the same issue of *Nature* entitled "Problems with Dating an Older Woman" in which he reviewed the controversy and the history of analysis. He pointed out that in the modern science of paleoanthropology understanding the geological context was assuming an importance in reconstructing human evolution equaled only by knowledge of the detailed anatomy of the fossils.

When Don Johanson had been a graduate student in the Omo he and Hank Wesselman had joked about inventing a board game for paleoanthropology. It would be something like Monopoly. You roll the dice and move around the board. Some of the options were "You have discovered an early hominid jaw; collect 100 kudos" or "You have been awarded a major research grant; collect $200,000." But there were also negatives, such as "Revolution in the research country; go to jail." In the real-life game of paleoanthropology, Johanson had now drawn the card that said "Your tuff has been redated, go back to Go." He had remarkably little to say about the redating at Hadar. Jim Aronson was contacted by a reporter and was quoted as saying that his date could be incorrect. Only Tim White continued the fight, sending a paper that attacked our faunal argument for the dating. But because he had apparently given up his attack on the primary geological evidence and the tuff correlations were generally accepted, his message was rendered largely academic.

## HOMINID ROOTS

But what was the real importance of the dates? Why was so much ink being spilled over this issue? The real issue was close to the hearts of all die-hard paleoanthropologists. It was *phylogeny*. This term also goes by the names of "genealogy," "lineage," "ancestry," or "roots." Many people other than professional paleoanthropologists can get worked up over issues like these, as anyone was has ever witnessed a discussion at the Daughters of the American Revolution of the question of George Washington's descent from English nobility will appreciate. And I have always thought that my own proximate origins in Virginia, where, as the saying goes, people are like potato plants (their most important part is underground), had predisposed me to paleoanthropology.

When people are figuring out their own genealogies, the dates that their ancestors lived are critically important to getting the branches to the family tree in the right order. This is why genealogists skulk around cemeteries recording birth and death dates and publishing them in books. For example, there have been a number of men named Thomas Boaz in my family, but only one of them, corresponding to dates in the early to mid-1700s is my ancestor. If I were investigating a letter written to a Thomas Boaz, thinking this was my ancestor, but the date was some time in the mid-1800s, then this Thomas Boaz could not be lineally related to me. The Hadar redating was similar.

*Australopithecus afarensis* was suggested as a lineal ancestor of all the later hominids by Johanson and White in 1978. This conclusion meant that on the phylogenetic tree of the hominids *A. afarensis* had to show all the correct morphology for an ancestor and it had to date early enough to serve as the ancestral population from which the later hominids arose. The redating of Hadar showed instead that the Hadar fossil was about the same age as South African *Australopithecus africanus*, a species that Tobias and I had already compared to the Hadar fossils. Thus the redating of Hadar was a big blow to the phylogeny proposed by White and Johanson.

The Hadar redating had backed Johanson and White's phylogeny into a corner, but another discovery in the Turkana basin knocked it onto the ropes. In 1986 Richard Leakey's team reported the discovery of a well-preserved robust australopithecine skull on the western side of Lake Turkana. The skull was catalogued as KNM WT 17000 but was nicknamed the Black Skull because of its unusually dark color, probably caused by manganese and iron in the sediment. This robust australopithecine skull had the clearly diagnostic features of a large bony crest running along the top of its head, massively built upper jaw, depressed nose region, flattened face, and heavy brow ridges. But most interestingly, it came from a level dated at 2.5 million years old, a very early date for robust australopithecines.

This date was not good for the Johanson-White scenario. They had proposed that *Australopithecus africanus* had given rise directly to the robust australopithecines, but this robust australopithecine was the same age or earlier than the dates that Johanson and White had ascribed to *africanus*. What to do? In a paper published in 1988, White proposed a more complex phylogeny, this time suggesting two separate lineages of robust australopithecines—one in South Africa descended from *A. africanus* and another one in East Africa that looked very similar but that had descended directly from *A. afarensis*.

To me the complexity seemed unnecessary. I was reminded of the ancient Egyptian astronomer Ptolemy, who modified his theory of planetary movement every time new contradictory observations were made. He had been convinced that planets had to move in perfect circles, and to accommodate the apparent wandering of planets he had had to impose circle upon circle to make his theory conform to observations. The increasing complexity of the theory really meant that there was something wrong with it. His theory eventually was replaced by an understanding that planets can move in ellipses of varying shape.

The Johanson-White phylogeny, I thought, was similarly committed to an ideal assumption—in this case that *Australopithecus afarensis* was ancestral to *Homo*, to the exclusion of *A. africanus*. In a detailed review of the *afarensis* question that I wrote in 1988 I suggested a simpler hypothesis. *Afarensis* could be kept as a valid species name because there were several small but significant morphological indications that it was more primitive than *africanus*. But it was still quite similar in overall morphological pattern to *africanus*—just a slightly earlier model. In my phylogeny (see endpapers), which was not very different from what I had suggested in 1979, *afarensis* was ancestral to *africanus*, which in turn was ancestral to the genus *Homo*. *Afarensis* is now dated to around 4 million years ago at other sites in the eastern Rift Valley, and it might even be represented by the earliest record of hominids we have, a jaw from Lothagam, Kenya, that is around 5.5 million years old. *Afarensis* evolved into *africanus* by around 3 million years ago. *Homo habilis* appeared at 2.5 million years ago, having originated from *A. africanus*. The origins of the robust australopithecines go back to 2.5 million years ago, but we are still unsure of their ancestry. They may have evolved from an *A. afarensis* group or even from *A. africanus*.

I feel that this phylogenetic arrangement still fits the data best, although other paleoanthropologists believe that other more complex arrangements are possible. Colleagues Todd Olson, Dean Falk, Yves Coppens, and Brigitte Senut, for example, believe that more than one species may be present at Hadar. One is australopithecine-like and the other is *Homo*-

like. My feeling, however, is that the simplest hypothesis to explain the facts is the best. In my opinion, the Hadar sample is most likely one population of hominids that had a relatively large degree of size difference between the sexes. But of course the nature of science is that hypotheses are always subject to refinement, refutation, or disproof. Continuing research at Hadar or other sites in the time period may reveal the presence of more than one species. If work were to resume at Laetoli, more complete specimens from there might show these hominids to be substantially different in skull and facial form from the Hadar hominids. Or continuing work at the South African cave sites might reveal that the perceived unique attributes of *afarensis* might be matched completely in an expanded *africanus* population, thus negating the necessity of the *afarensis* name.

The furor over Lucy died down, but papers on *Australopithecus afarensis* dominated the professional meetings and the pages of the journals during the 1980s. When Ethiopia called a temporary moratorium on paleoanthropological research, Don Johanson and Tim White moved operations to Olduvai Gorge, displacing Mary Leakey, who had worked there since the 1930s with her husband, Louis. Johanson and White discovered a partial skeleton of *Homo habilis*, Olduvai Hominid 62, and then moved back to Ethiopia when the moratorium was lifted. Recent fieldwork at Hadar has turned up some more fossils of *afarensis* but, not surprisingly, the press coverage and the professional interest have not matched the initial reception of Lucy.

Hadar will always be an important part of the story of hominid evolution, but it can only relate to what happened in northern Ethiopia between 2.8 and 3.4 million years ago. Other questions in hominid origins now seemed bigger and more important to answer than those that can be answered at Hadar.

I was particularly interested in the overall theater of action in hominid evolution. This theater is unique, with many levels of stages. The topmost stage is well-lit, and the curtain is fully drawn on the modern world with all its multifarious peoples populating the whole globe. But the lower stages, representing successively earlier times in human evolution, are more dimly lit, and sometimes with the curtains only partially opened. The famous East African Rift Valley sites of Omo, Turkana, Olduvai, and Hadar provide us with only a partial view of the edge of the stage. The South African australopithecine cave sites open up the curtain a little more and provide another, slightly wider view, but they are on the same stage. The geological ages of these sites are the same as those of East Africa. I was interested in opening the curtain on new stages, ones that until now had been unlit and unviewed.

Early hominid evolution was limited to the African continent, a fact that we know from extensive research on the early sites of Eurasia and the Western Hemisphere that has yielded no trace of the earliest hominids. But most of Africa is totally unknown. It remains the dark continent, a succession of stages of human evolution unlit by research. I sought to investigate the parts of Africa that paleoanthropologists had not yet succeeded in opening up. I also wanted to find sites that were earlier than those of eastern and southern Africa in order to trace hominid origins back to the beginning.

I decided to start with North Africa.

# 5

# Rivers of Sand, Animals of Stone

Benghazi, Libya, April 1976. Stuffed into a Toyota Landcruiser, we were on the road into the desert by 5:00 A.M. We were going to Sahabi, a reportedly rich fossil site abandoned by Italian researchers in the 1930s. Sahabi lay on the ancient caravan route that stretched south from Benghazi into the desert. Our tour guide was the dean of the faculty of science of the former University of Libya at Benghazi, now renamed Garyounis University, after a nearby army post where Muammar Qaddafi had once been stationed. The group included various faculty members from Benghazi and myself. As by far the youngest, I sat in the back of the Landcruiser, despite the protests of my hosts. As we were leaving Benghazi, the dean, an ebullient and likable man, ordered the driver to stop, and we had a predawn roadside breakfast of eggs cooked between layers of flat bread swirled like pizza dough into hot oil. Just the sort of cholesterol-rich breakfast one needs for trekking through the desert with nothing else to eat all day.

Driving south at 100 kilometers an hour along a surprisingly smooth modern paved road elevated several meters above the shifting desert sands, I reflected on the challenges of Sahabi. A visiting British paleontologist who years earlier had worked at a fossil site in central Libya had come to Berkeley for a lecture. I asked him about Sahabi. Had he ever been there? What were the field conditions like? His answers were daunting. No, he had never been to Sahabi. He had avoided the entire area because it was still heavily mined. German Field Marshall Rommel's Afrika Korps had been headquartered for much of the war in a desert town just north of Sahabi, and the tank battlefields to the south had been one of the most active theaters of action. In addition, he had been near Sahabi when he was captured by desert nomads who had threatened to kill him. Only by reciting several lines from the Koran had he been spared. He commended these lines to me to memorize for future protection. I was reminded of the story of the British explorer who had found the fabled city of Timbuktu but had been murdered on his way back across the Sahara by his Muslim guides when he refused to convert to Islam.

I wasn't sure how serious the threat of desert nomads was. Still, I had memorized the relevant lines of Arabic.

I immediately went to the university ROTC office, not a particularly popular place on the Berkeley campus during the early 1970s. I was surprised that it still existed after the furor of the antiwar movement during the 1960s and the virulence of the then-current antimilitary feeling. I had wanted information on the war in North Africa, specifically land mines. The commandant of the program had served in North Africa during World War II, and he told me off the top of his head about the land mines. Almost all of them were antitank mines. A person walking over them would not set them off. He was not aware that any "Bouncing Betty" antipersonnel mines had been deployed in North Africa. These devices were spring-loaded so that after a person walked over them they bounced out of the sand to blow up above-ground, spraying shrapnel in all directions. Also, there had not been any of the wooden land mines that are undetectable by metal detectors. Later, from a colleague in Germany, I received a map of the oil fields in north central Libya that indicated that the Sahabi area recently had been cleared of land mines. This discovery had been enough to reduce the risk to an acceptable level for me and the others whom I would take into the field with me.

## DISCOVERY OF THE FIRST FOSSILS

The scenery changed from an arid Mediterranean-type vegetation, not unlike Greece, to fewer and fewer patches of green as we went south. The few buildings and occasional towns gave way to emptiness. Nothing lay ahead of us to the south but the flatness of the full desert.

We arrived at the site of the old Sahabi fort, which had been first a Roman, then a Byzantine, then a Turkish, and finally an Italian colonial installation guarding the ancient caravan routes across the Sahara. The walls were mostly destroyed by World War II tank fire, but some of the arches and foundations remained. We started out across the desert with only some of the old imprecise Italian paleontological reports to guide us.

I walked through the desert warily, as one does when exploring new, unfamiliar, and potentially dangerous territory. This environment was totally different from any fossil site I had ever encountered. There was sand everywhere—not the soft, white, beach variety, but a hard, brownish, gravelly pavement winnowed by constant winds. How, I wondered, could anyone find fossils here when all the rocks were covered by sand? How could you ever plot the location of sites on aerial photographs in this treeless, featureless landscape? How could fieldworkers find their way? Had all the land mines that peppered this terrain during World War II actually been removed?

As my first day in the desert was drawing to an end, I was beginning to feel concerned. We had spent most of the day walking around the Sahabi desert without a sign of bones. I had not expected any significant discoveries, but I needed a sign, some indication that fossils were here. Nothing. Another hour of walking in the desert parallel to the road as the Landcruiser kept pace with us. Slowly the others gave up and walked back to the car. I found myself alone. The sun began to slant across the sand, and I estimated there was maybe another hour of daylight left. I was beginning to feel very tired. And I was thirsty. Although it was spring and the sun was not particularly hot, the constant wind dries you out.

Then I saw it. Something familiar in an unexpected setting, like an American pop song suddenly coming over the radio in Cairo or a friend's picture in the newspaper. *Lates niloticus*, the Nile perch. There it was. I'd recognize the dimply bone surface of the skull anywhere. They were all over the Omo.

As I was running to the Landcruiser I had to smile at myself. Here I was elated at finding a small fragment of a fish when only a few years before I had been secretly disdainful of the hundreds of mammal fossils and isolated hominid teeth we had been excavating at Omo. I had wanted more complete remains, like skulls. But this situation was different. If there were fish fossils, so common at Omo, there ought to be many other fossils associated with them here at Sahabi. The processes of fossilization should be the same. And Sahabi was so much older than Omo. Who knew what we could find?

My long-suffering Libyan friends in the Landcruiser were genuinely pleased that I had found what I was looking for. I must have been a rather forlorn sight walking through the desert and looking down at the sand in an apparent attitude of dejection. Everyone piled out of the Landcruiser for another look in the waning light. We found more bones: crocodiles, other fish, and many unidentifiable bone fragments. Almost all had been broken and smoothed by the blowing sand grains. Then Ali El-Arnauti, a geologist who was to become one of my closest collaborators and friends over the next several years, motioned to me to come over. He had found a headless skeleton the size of a man, but the ribs were too thick and the limbs were too short. I guessed that it was a pig. We had to leave it because it was now almost dark and I didn't know how to record the location. I photographed it and marked the kilometers on the Landcruiser odometer. We found the place the next year, but the entire topography had been changed by the winds. There were new dunes and no bones where there had been before, and new depressions with bones where there had been dunes before. We never found the skeleton again.

On the drive back to Benghazi our mood was upbeat. One of the group was a professor of classical archaeology, and he gave us a history of the Nile perch in ancient Egyptian theology. An alternative name for Luxor had been Latopolis, named after the perch god. Ali, the dean, and I talked about the future of an international project to investigate Sahabi. As the others drifted off to sleep I mentally ticked off a status report. There were fossils, but the dunes obscured a lot of the exposure; recording and relocation of sites was going to present a challenge; the exposures were limited and the geological study would be long and arduous; and logistics would be a nightmare. The threat of land mines was hardly the worst of it.

## SEARCHING FOR THE MISSING LINK ON A NEW STAGE

Sahabi was an entirely different undertaking than the East and South African sites from which early hominid fossils had been found. First, there was the contrast between the current environment and the world from which the fossils had come. The environments of the Serengeti Plain in Tanzania and Kruger National Park in South Africa, for example, were quite similar overall to the famous fossil sites at Olduvai and the South African cave sites. It was not difficult to imagine yourself as an early hominid walking across the savanna as you finished work on one of the excavations at Omo. But Sahabi required a real excursion of the mind. There was nothing left of the former lush vegetation that must have supported the elephants, the antelopes, and the crocodiles that the Italian researchers had found there. Nothing but sand. The entire paleoenvironment would have to be reconstructed from scratch, using every piece of available geological and paleontological evidence.

Sahabi was much earlier than all of the well-known East and South African sites. The animals would be different, and the hominids—if there even had been hominids there at that time—would be certainly different from Lucy and her kind, generally conceded to be the first well-known hominids in the fossil record. Sahabi was thought at the time to be around 7 million years old. Hadar and Laetoli are dated to 2.8 to 3.7 million years ago.

I was particularly interested in Sahabi because of its age; its geology, which promised the recovery of complete fossils; and its location outside the well-trodden paths of eastern and southern Africa. The end of the Miocene Epoch and the beginning of the Pliocene, to which Sahabi dated, was the focal point for the investigations of the evolutionary split between the African apes, our closest relatives, and our lineage, the hominids. Now that I knew where in the desert to begin to look for fossils, we had a place to start.

---

I had once before made a foray into the desert to look for fossils, but that had been in northern Kenya. There were a few small sites near this time period in the Lake Baringo basin that Louis Leakey and British geologist Bill Bishop had discovered, but subsequent expeditions had turned up only very fragmentary hominid or near-hominid remains. They were too fragmentary to allow a definite identification. A specimen from the 5-to-6-million-year-old site of Lothagam was the most complete, but it was only a piece of jaw with a molar tooth that looked hominid. The height of the jaw was greater than that of an ape and the molar was flat and low-crowned, like a hominid. The Lothagam jaw was considered the first evidence of hominids in the entire fossil record. Another nearby site in Kenya, Lukeino, had yielded an even earlier but even more fragmentary fossil document, a single molar tooth dated to about 7 million years ago. Some anthropologists thought this specimen was a hominid, too, but many others preferred to reserve judgment.

Before going to Sahabi, I had read all the reports on the geology and paleontology of the Baringo sites. I had also driven to the remote parts of northern Kenya—the Baringo basin, the Suguta valley—to see the terrain for myself. It was a low-lying part of the East African Rift Valley, too hot and arid for people, except for the occasional nomadic camel drivers. I had once stopped to ask directions from one of these men, but he understood neither Swahili nor any European language that I tried. He may have answered in Arabic, a language with which I was to become well acquainted only several years later.

It was a severe but starkly beautiful landscape. In some parts the land was strewn with great dark boulders spewed out by ancient volcanoes. They reached up to the Landrover's window and were too close together to allow the car to pass. They had to be negotiated on foot. Outside the Landrover the midday heat took your breath away. The black volcanic rock absorbed the sun's radiation and created a layer of superheated air that clung to the ground surface. On one of my solitary surveying forays the heat had gotten to me. I found myself walking on and on, dazed and repeating to myself disjointed stanzas of a half-remembered poem by Byron: "O the dreary, dreary moorland. O the barren, barren shore." Only the sudden unexpected startling of a spring hare out of its hiding place brought me back to my senses sufficiently to make me realize that I was wandering. I became angry. In environments like this, losing even a little bit of control, even slight errors in judgment, could be fatal. I managed to find my way back to the Landrover, drink some water, and cool off. This experience in the Suguta valley taught me a valuable lesson about respecting overbearing heat—say, over 120 degrees—and the physical limitations associated with it. I had learned something that would stand me in good stead for the Sahara.

I also thought that I understood why only fragmentary bones had come from the Baringo basin sites. The fossil deposits were what geologists called high-energy deposits. They were large-diameter gravels, cobbles, and boulders moved by rapidly flowing, even torrential, water. Another clue was that many of the sediments were poorly sorted. There was a mixture of coarse sediments along with fine-grained sand and silt, indicating that the sediments with the bones had been laid down rapidly. The water had carried away everything in its path until it had rapidly dropped its sediment load, including the bones. A more gradual process of sedimentation would have allowed the different-sized sediment particles to drop out of solution as the speed of the water's flow subsided. I believed that the rapidly moving water and the suspended gravel and cobbles had broken up and rounded off all of the fossils that had been deposited in the Baringo sites. Single teeth and isolated jaw fragments were among the densest and hardest skeletal parts, the ones that you would expect to have best weathered this abrasion and breakage, and they were the pieces found at Baringo. Omo had shown me that single teeth and isolated jaws, even large numbers of them, were not going to answer the burning questions. I decided that I needed a site in which there were well-sorted and low-energy deposits so that more complete remains could be found.

The sediments at Sahabi had been reported by the Italian geologists to be fine-grained silty sands. And the most famous discovery at Sahabi had been a complete skeleton of a four-tusked elephant (two tusks in the skull like modern elephants and two tusks protruding from the lower jaw). Carlo Petrocchi, the leader of the Italian expedition, had named this remarkable animal *Stegotetrabelodon lybicus* in 1934. The completeness of the skeleton and our own discovery of an articulated headless skeleton on our first visit to Sahabi had convinced me that we were on the right track.

Andrew Hill of Yale University, a former student of Bill Bishop's, continued to investigate the sites of the Lake Baringo basin and found a hominid jaw from the site of Tabarin, dating to some 4 million years ago. Hill has ascribed the specimen to *Australopithecus afarensis*. The antecedents of *afarensis*, however, still elude us.

## THE AGE OF SAHABI

The Italian paleontologists who had worked at Sahabi in the 1930s had ascribed the site to the Miocene age. But how old was Sahabi really and how confident were the dates? The initial estimates had been made on the basis of comparisons of the fossil animals found at Sahabi with those in Europe, ascribed to the Miocene, Pliocene, or Pleistocene epochs, and local stages within those epochs. The species that the Italians had found at Sahabi looked different from European species but bore the most similarity to those of the late Miocene Epoch. There certainly had been no four-tusked elephants running around with Pleistocene Neanderthals, or in the preceding European Pliocene for that matter. The Sahabi fossil antelopes and pigs seemed to suggest the same conclusion.

After the publication of the initial findings of the Italian expedition, World War II intervened and Carlo Petrocchi died. No one in Italy took an interest in going back to Libya after the war. The geologist who had first brought scientific attention to Sahabi as well as being responsible for discovering the first of Libya's vast oil reserves, Ardito Desio, was now the grand old man of Italian exploration. He was leading exploratory expeditions to the Antarctic and mountaineering parties to the Himalayas. Sahabi languished.

Meanwhile geological dating techniques were undergoing a revolution. Willard Libby's carbon-14 technique worked for dating back to around 50,000 years. In the East African early hominid sites the absolute dating method of choice had become potassium-argon, useful in assessing much older dates. The method had worked wonderfully in the East African Rift Valley deposits where fossil levels were interlayered with volcanic ashes and lavas, high in potassium content.

Sahabi was a different story. Far from any volcanoes, it had no potassium-rich rocks and the site was much too old for carbon-14. But the old techniques could be applied with a twist. If a species found at Sahabi could be located in other sites that did have absolute dates, those dates could provide an age estimate for Sahabi—sort of relative absolute dating. The first paleontologist to apply this technique to Sahabi was Vincent Maglio, who in 1970 published a monograph on African elephant evolution. Maglio had found other *Stegotetrabelodon*s in the East African Baringo sites. In his study of the type specimen from Sahabi, still on display in the Tripoli Museum of Natural History, he had traveled to Libya and brought Sahabi back to the attention of the scientific community. From comparison with absolute dated sites in East Africa, Maglio thought Sahabi was about 7 million years old. This conclusion meant that the site was late Miocene in age in traditional terminology.

This time period was perfect for me. I wanted a site that could throw light on the ancestors of the African apes and the hominids or on either one of the lineages after their split. Recent happenings in the worlds of molecular evolution and paleoanthropology had changed the way that scientists were

viewing the timing of this evolutionary split. Molecular anthropologists Vince Sarich and Allan Wilson had been maintaining
since the mid-1960s that the split could not be more than four
or five million years ago. With some prodding they and other
molecular evolutionists, such as Jack Cronin, a graduate school
friend of mine then at Harvard, would push their data to an
outside statistical possibility of an 8-million-year-old split, but
no earlier. This steadfast confidence in what appeared to mor-
phologists to be very insubstantial stuff—molecules—amused
some and angered others. I had been in the former category,
although I had had a good genetics background as an undergraduate. After taking a graduate reading course with Sarich as
a graduate student and after endless discussions with Jack
Cronin, my amusement gradually turned to bemusement. Their
data, assumptions, and cross-checks seemed to bear out their
findings. Why couldn't they be right? What was the fossil and
dating evidence to the contrary?

## THE LATE DIVERGENCE HYPOTHESIS

In the mid-1970s the paleoanthropological status quo on the
Miocene beginnings of the hominids derived primarily from
the work of Elwyn Simons and David Pilbeam, on the one
hand, and Louis S. B. Leakey, on the other. This position is
now known as the early divergence hypothesis because it
argues that the evolutionary splits of the hominids and apes
occurred some 15 to 20 million years ago. Simons and Pilbeam
had reviewed and revised the naming of all the fossil hominoids
in the world. Their interpretations were that the great apes—
the chimpanzee, gorilla, and orangutan—had split off early in
the Miocene Epoch. In their scheme, the African early
Miocene fossils were sorted by size; the big one was a species
ancestral to the gorilla, the medium-sized one migrated to Asia
and was ancestral to the orangutan, and the small species was
likely ancestral to the chimps. The middle Miocene genus
*Ramapithecus*, found first in India and later in Pakistan, was
fingered as the most probable hominid ancestor. This would
have put the ape–human split at about 15 million years.

Leakey's evidence came into the story in the form of an ape

jaw and partial bony face found at the middle Miocene site of Fort Ternan, Kenya. He had named them *Kenyapithecus*. Simons and Pilbeam had sunk Leakey's name to *Ramapithecus*, indicating that they believed the two fossils to represent the same species of animal. Leakey, of course, had not accepted this opinion. He traced *Kenyapithecus* back to the early Miocene, holding that hominids had diverged from the ancestors of the living African apes even by this time, back about 18 or 20 million years ago.

Then something unexpected happened. A new fossil skull turned up in Pakistan. It was a close relative of *Ramapithecus*, and was ascribed to *Sivapithecus*. What was important was that the skull, dated by absolute methods to between 8 and 9 million years ago, looked very much like an orangutan. Jack Cronin jumped on this fact. In a paper published in the British journal *Nature*, he and paleontologist Peter Andrews pointed out that if *Ramapithecus* (and possibly the fragmentary *Kenyapithecus*) were ancestral to the orangutan, then there was no believable fossil evidence for the early divergence of the hominids. In other words, there was no paleontological impediment to accepting the molecular arguments for the timing of the ape–hominid split, a hypothesis now dubbed the late divergence hypothesis. A sheepish David Pilbeam, whose expedition had found the skull, admitted that he had been wrong about the *Ramapithecus* and *Sivapithecus* leading to the hominids. Not only did an early divergence seem unlikely, but an Asian origin for the hominids was also ruled out by the new evidence.

## THE GEOGRAPHY OF HOMINID ORIGINS

By 1976, when I had read Maglio's elephant monograph and written to the Libyan authorities about reinitiating research at Sahabi, 7 million years old in Africa seemed a very good time and place to be looking for fossils potentially ancestral to our own lineage.

But why North Africa? Beginning in 1925, when Raymond Dart had published the discovery of the South African "man-ape" *Australopithecus*, everyone thought of sub-Saharan Africa when they thought of hominid origins. But there was no intrinsic reason that this should be so. There were caves in South Africa that had served as bone collection points, and in East Africa there were the subsiding basins of the Rift Valley, covering up the bones as water and volcanoes laid down sediments. The logic that there had been hominids where their bones were found was inescapable, but the converse, that there had been no hominids where there was no knowledge of their bones did not follow. Only if intensive searching had been carried out and there had been no discoveries did this assumption make sense.

I knew that no one had seriously looked in North Africa for the earliest remains of hominids, but they had looked and made discoveries on both sides of the question: early ape ancestors and much later *Homo sapiens* fossils. Elwyn Simons of Duke University had researched the rich primate fossil beds of the Fayum site in Egypt, the most important site in the world for a picture of Oligocene ape and monkey ancestors 36 to 40 million years ago. And French researchers, particularly the late Camille Arambourg of Paris, had made important discoveries of archaic *Homo sapiens* and *Homo erectus* dating back several thousands of years in Algeria and Morocco. The British archaeologist Charles McBurney had excavated the cave of Haua Fteah on the Libyan Barbary Coast and had discovered a human mandible comparable in age to the European Neanderthals. There was only one record of any fossil that possibly could have been be an early hominid. In the 1960s, Yves Coppens of the Musée de l'Homme in Paris had found a skull near Lake Chad that he had named *Tchadanthropus*. I had studied the specimen in Paris. Blowing sands in the Chadian desert had scoured off the outer few centimeters of the entire skull, including the all-important surface detail. The outer rims of the orbits were gone, giving the skull a look of wide-eyed bewilderment. It might represent an early hominid, but nobody would ever be able to tell from this specimen. Sahabi, with its promise of low-energy fossil-bearing sediments of the right age, might hold the answer to the question of African hominid origins, but this time from a very different part of the continent— not just north, but also west.

## MOUNTING THE EXPEDITION

My first job in Berkeley after returning from Libya was to formulate a plan of attack. After developing my slides, I gave my professor and mentor Clark Howell a private view of the foray into the desert. He accepted my assessment that Sahabi showed a lot of promise. After earlier reports about land mines, rapacious nomads, and unsettled politics, he seemed somewhat surprised that things had gone so smoothly in Libya. Still he showed little enthusiasm for going to Sahabi himself, although he was very interested in studying the fossils when they came out.

Clark agreed to help me obtain funding for the Sahabi expedition. It was not going to be easy to raise money for a scientific venture in a country hostile to the United States and at a site abandoned for forty years whose potential was still unproven. We put together a prospectus, a fancier-than-usual research proposal with photographs and a bound cover. We reproduced a commemorative Libyan postage stamp with the Sahabi *Stegotetrabelodon* on the front. Clark gave copies of our prospectus to various potential backers with whom he had contacts through the L. S. B. Leakey Foundation, a nonprofit funding organization for human evolutionary research based in California.

Gordon Getty, chairman of the Leakey Foundation Board, expressed an interest, and we met with him in his San Francisco home, an Italianate palazzo. When we arrived, Getty had been in his study working on calculus problems. The son of oil magnate J. Paul Getty, Gordon was now chairman of Getty Oil and had recently been ranked number one on *Forbes* magazine's scale of individual net worth. Yet in the business community it was whispered that he was more preoccupied with music, art, and science than with money and that he fashioned himself a modern Medici. From what I could see, that description seemed pretty apt.

Getty agreed to put up a portion of the initial funding for the Sahabi expedition, enough to buy a Landrover and field equipment. He agreed to help us find other funds through the

Leakey Foundation. Foundation trustee Leighton Wilkie agreed to donate the rest of what we needed to get a small geological and paleontological survey team into the field. I put in an order for a Landrover at the British Leyland factory in Solihull, England, to be shipped and in Libya by December 1977.

In the meantime I had finished my Ph.D. thesis and accepted my first job, lecturer in anthropology at UCLA. I stayed in contact with the Libyans, and I began to put together our research team. Geologists Jean de Heinzelin of the University of Gent, Belgium, and Frank Brown of the University of Utah, both of whom had worked with me in Ethiopia, agreed to go with me to Sahabi in the spring for a reconnaissance.

Problems developed at the Landrover factory. A strike had delayed all orders, and delivery could not be guaranteed until mid-1978. This was too late. We had to have a vehicle in Libya for the reconnaissance in the spring, when all of us had a few weeks between sessions. The best the factory could do was to confirm that the Landrover could be manufactured by the end of December, but the delivery could take months. I wrote to have the delivery price deducted from our invoice and found an inexpensive excursion ticket to London. For the cost of shipping the Landrover to Libya I could fly to England, drive the Landrover to Naples, and take a ferry to Benghazi over Christmas vacation. This plan would insure that the vehicle would be there when we needed it.

With the ending of the school quarter at UCLA, I set off to London. I arrived early in the morning and immediately went by taxi to the Landrover factory, where they were just opening up. I paid the last of the bill, arranged for temporary registration and insurance, picked up the Landrover, and set off for Dover. It was cold, and I was suddenly glad that the factory, in their haste to get my vehicle ready, had failed to remove the heater as I had requested. "Why would you need a heater in the Sahara?" I had thought. The sea on the ferry ride to Calais was rough, and when I arrived it was raining. I headed for Belgium, and I hoped to reach Brussels that night. However, I discovered that the windshield wipers didn't work. My progress out of Calais was slow because I had to wipe off the rain with my left hand out of the window as I negotiated narrow, twisting streets. It was now dark. As soon as I reached the freeway to Brussels, visibility was better and eventually the rain stopped. But I was getting very tired. I reached Brussels at about 11:00 P.M. and phoned Jean de Heinzelin. Luckily he was still up and waiting for me. After wandering around a bit I found his house, had one of Jean's excellent Trappist Belgian beers, and went to sleep.

I had intended to set off early the next morning, but Jean convinced me to stay one more day in Brussels, have the Landrover wipers fixed, and rest up a bit before the drive across Europe. We got the Brussels Landrover dealership to free up the wipers and run a general check, while Jean contacted the Automobile Association for the best route to Naples and road conditions. I would have to cross the Alps, and depending on weather conditions roads could be closed on short notice. This information proved very useful. The route that I was to take led through Luxembourg, southwestern Germany, eastern France, Switzerland, and Italy. It was threatening to snow, but the best route through the Alps was the Pass Grand St. Bernard.

I set off very early in the morning because I wanted to beat Brussels rush hour traffic. I was in Luxembourg by nine, Germany by ten, France by noon, and Switzerland by late afternoon. It began to snow. I kept going into the Alps, however; I did not want them to close the road on me. As I reached the highest sections of the road there were blinking warning signs saying that the road was going to be closed. I kept going, using the Landrover's high range four-wheel drive for the first time. This scenario was certainly different from the one I had imagined for my first four-wheel-drive excursion in this particular Landrover, which I had ordered in sand color for the Sahara. I made it over the top of the Alps and entered Italy. After a long day's drive I checked in at a small hotel north of Milan.

Two mornings later I was in Naples negotiating with the shipping company for passage across the Mediterranean. Nobody spoke English, so I used Spanish, with what I thought sounded like an Italian accent. It worked pretty well. I managed to find out that the ferries only ran once a week to Benghazi and I had just missed one. But the shipping company official assured me that I could hand the vehicle over to them if I was in a hurry and they would ship it and deliver it for me. No problem. It was done all the time. I was dubious, but I took all the forms, bought an Italian dictionary, and translated the fine print. It looked legitimate, and I didn't have the time to wait in Naples a week to get the ferry to Benghazi. By then I had to be back in Los Angeles for the start of classes. I called the University Research Center in Benghazi and got all the details on location for delivery, telephone numbers, and people to contact. I then went back to the shipping office, arranged for the maximum insurance, signed the papers, and handed over the keys. It was the best I could do and I could only hope that all the time and effort that had gone into getting this Landrover this far would not have been in vain. I found out by letter a few weeks later that it had arrived safely. I had no way of knowing the many hundreds of hours and the thousands of dollars that would be invested in this vehicle over the next fifteen years on my various African expeditions.

———————

Three months later, when I found myself again on the way to Libya, I was determined to make it. Jean de Heinzelin met me in Brussels. We were to fly the next day on the same flight to Rome, where we were to meet Frank Brown, who had taken the opportunity to travel through southern France to investigate some volcanic deposits. Just as we were sitting down to dinner at Jean's house, the phone rang. It was the American consulate in Lyon, France, with doubly bad news. Frank had been waylaid and robbed by an escaped convict in southern France, and a border war had just erupted between Libya and Egypt. They suggested that we call off the trip. We told them we would call them back in an hour.

I was against calling off the trip. It had been difficult scraping together the funds for this pilot study, and we had already spent most of the grant from the Leakey Foundation for the Landrover and for the tickets to Libya. If we now had to turn back with no results and almost all the money spent, it would be very difficult to raise more. Even if Frank couldn't go, Jean and I decided to proceed to Rome and contact the Libyans from there as to conditions in the country. We called the consulate and told them that we would leave as scheduled for Rome and that Frank should meet us if he could.

When we arrived in Rome all the flights to Libya were canceled because of the war. We checked into a hotel and waited. We made use of the time by going to the University of Rome Geology Museum to study the fossil specimens still housed there from Sahabi. Frank showed up in Rome the second day. He had made the mistake of picking up a hitchhiker in France who had just escaped from prison. The man had stolen Frank's rented car with all his luggage when Frank had made a geological stop to collect some rock samples. After reporting the incident to the police, he had been on his way to Rome by train when he had seen the man on the same train, carrying Frank's suitcase with him! The man jumped off the train when Frank saw him and Frank jumped after him, tackling him. The police sorted the story out, returned Frank's luggage, which included almost everything that had been stolen, and took the escaped convict back into custody. So Frank was in Rome, ready to forge ahead into the Sahara, none the worse for wear except for a slight bruise on his nose where the convict had punched him.

The next day the Libyans negotiated a truce with the Egyptians over their shared border and the flights resumed. We boarded an Alitalia flight and finally were on the last leg of the trip to Libya. All we had really done was just get there. All of our real work lay ahead of us. But as is frequently the case in undertaking fieldwork in Africa, getting equipment and scientists to the field site is a major challenge and I somehow felt that we had scored a success already.

Our arrival in Libya coincided with the Islamic holy month of fasting known as Ramadan. No one can eat or drink during the daylight hours, and everyone at the university was working at half speed. But our arrangements proceeded smoothly, we were reunited with our Landrover, and we soon were off to the desert with two assistants provided by the university. We were given a Toyota Landcruiser with driver on loan as a backup vehicle and a house in Agedabia, a town on the edge of the desert, about an hour's drive from Sahabi. We stayed there for two weeks and made daily forays into the desert around Sahabi. Because our Libyan assistants were fasting and could not drink water during the day, they would prepare large containers of water in the morning before sunrise as we left for the desert, think about them all day, and then, back at the house, down them in one long gulp as the evening muezzin signaled the Islamic evening prayers. Their wives prepared wonderful soups, specially made during Ramadan because they took all day to simmer. We never saw the wives however, and only expressed our appreciation via their husbands. To have directly thanked the women would have been impolite. This custom particularly distressed Jean, who never became used to the segregation of the sexes in Islam and said so.

We started at the Sahabi fort and, using the old Italian maps as well as we could, tried to relocate all the former fossil collecting areas. We found the site of excavation of the *Stegotetrabelodon*, the old air field (the *campo d'aviazione*), and the other localities. Frank plotted them at a different scale and discovered a pattern. The fossil localities were in two wide swaths through the desert. We hypothesized that there might have been two ancient channels in which the fossils had been deposited. But as we surveyed, we realized that there was a much more prosaic explanation. All the fossil sites were on the sides of old roads through the desert, which forked as they went to the airfield and to the fort. Apparently soldiers and paleontologists had kept fairly closely to the established roadways. This discovery made us think that the fossiliferous exposures might be more extensive than the Italian researchers had realized, because they apparently had not ventured too far afield.

Jean and Frank started measuring and drawing stratigraphic profiles at areas of outcrop around the desert. We established a system of naming and numbering the localities using P (for point) numbers. Jean began compiling a detailed map orienting from the Sahabi Fort, a clearly defined landmark in the desert. We found a large number of fossil bones at locality P4, but we collected only a few specimens. The rest would have to wait until the upcoming field season in the summer.

As soon as I returned to UCLA I began work on the grant proposal to the National Geographic Society for the summer expedition to Sahabi. Grant proposals for totally new projects in areas of unknown productivity and with unknown logistical problems are always somewhat dicey. These areas also, of course, are potentially the most rewarding because you are exploring new terrain. The challenge was to make the project sound different enough to be interesting but not so different that there would be too many questions about how the project would be carried out. I tried to be realistic about the prospects of the research but at the same time communicate the team's enthusiasm for what we expected to find. The proposal would be sent by National Geographic to a number of peer reviewers, scientists who would pass scientific judgment on the merits of the proposal. The Research and Exploration Committee of the society would then take all the commentary into consideration and make their final determination on the grant. I found out several weeks later that the committee had given us a good review and that National Geographic had decided to gamble on the Sahabi Project and would fund our first full field season.

## DISCOVERY OF THE FIRST PRIMATES

The field season, our first large-scale investigation into the paleontology of Sahabi, started in the summer of 1978, two years after I had begun the planning. I thought we were ready. To cope with the extreme reflected glare of the sun off the desert sand I had bought some welder's glasses for surveying. To clear sand from sites for excavating I had bought a portable backpack leaf-blowing machine that I had seen workers using in affluent sections of Los Angeles. To cope with the problem of not being able to excavate in the heat of the day I worked out a portable generator and flood light system that would allow us to excavate at night when it was cooler. We had all the standard expeditionary equipment such as tents, cooking utensils, and excavation tools, and would buy additional supplies in the Benghazi markets. One of the essentials was head cloths, those heavy black-and-white or red-and-white tasseled cloths tied around the head in the desert manner. At first I thought this headgear was a bit too romantic, but after losing field hats to the desert wind, getting bad sunburn on the back of my neck, and having no eye protection in sandstorms, I found it very practical indeed. My welding goggles and sand blower did not work so well. Although the goggles cut down glare effectively in the noonday sun, several tests showed that it was still impossible to see a dark object the size of a quarter on the desert sand. And the sand blower worked well but was too heavy and cumbersome to cart around the desert. We used it a few times, but then it sat unused in a corner for weeks. Two members of the project quietly sold it to the Sahabi Petrol Station for gas and then told me about it later.

The most important element of my preparations had been the assembling of the research team. Two principal members of the team were from the Libyan side. Geologist Ali El-Arnauti had gotten his doctorate at the University of Muenster in Germany, and he was going to work with Jean de Heinzelin on the stratigraphy of Sahabi. Paleontologist Abdul Wahid Gaziry (Wahid for short), who had his doctorate from the University of Hamburg in Germany, would help me with the overall organization and would undertake the study of the fossil elephants. Both Ali and Wahid were in the Department of Earth Sciences at Garyounis University, Benghazi. Although both were native Libyans, they had gone to secondary school, university, and graduate school in Germany. Ali was the practical, responsible chairman of the department, and his approach was methodical and quite Western. Wahid was the mercurial intellectual, passionately interested in classic American jazz, fossil elephants, and ladies—not necessarily in that order. Jean got on famously with Ali, but he spoke disapprovingly of Wahid as being "too oriental." He had to admit, however, that Wahid's scientific papers were very good. Aspects of Ali's, Wahid's, and Jean's personalities resonated well with mine, although at first we would have seemed very disparate.

The American contingent of the field team was composed of

my former wife Dorothy, then a graduate student at Berkeley; Ray Bernor, a recent Ph.D. from UCLA; and a UCLA graduate student from Greece, Paris Pavlakis. Wahid had arranged for a contingent of German-based paleontologists to join us in the field. This group included Hamburg University Professor Ulrich Lehmann, who with Herbert Thomas of Paris was going to study the Sahabi antelopes; Ahad Salah, an Afghani specializing in invertebrate paleontology from Hamburg; and Herr Lierl, a geological technician.

The research specialists that we had assembled for what we were calling the International Sahabi Research Project were among the best in the world: Basil Cooke from Canada for the fossil pigs, Daryl Domning of Washington for the sea cows, Heinz Tobien of Mainz for the horses, Clark Howell at Berkeley for the carnivores, Roger Dechamps of Belgium for the fossil wood, Jean Gaudant of Paris for the fish, Jens Munthe then of Denver for the small mammals, and others. Most of these workers would never be in the field with us but would depend on our field team to collect the specimens, label them and record their contexts accurately, and figure out the stratigraphic and age relationships.

We entered all the specimens in our computerized catalog, which worked like an inventory list of a large department store. We stored the specimen number; its locality, which could tie a specimen to a spot in the desert no larger than a tennis court; its stratigraphic level; the name of the biological group the fossil represented; the body part that the fossil represented; who identified it and when; where the specimen was now; and some other more minor points of information. The computer kept the records in numerical order, but when we wanted to send a list of the pigs, for example, to Dr. Cooke we could sort out all the various records relating to this group and send him a complete and up-to-date list. The master list also served the purpose of keeping track of where the various specimens were after they had been sent around the world for identification. Eventually all of the specimens would come back to Libya for permanent storage in the Geology Museum at the university.

As we went out in survey crews, we collected increasing

numbers of fossils. We confirmed that the best fossil localities were not around the Sahabi fort but to the east in a semicircular twenty-five-kilometer arc around a big low-lying depression in the desert known as a sebkha. We established more and more localities, plotting them carefully on de Heinzelin's and El-Arnauti's map, photographing each one, and driving in wooden posts as an added measure to insure we could find them again. The next year we were able to make aerial photographs of the entire area, and we also plotted the localities on these.

One afternoon several weeks into the field season Wahid came into the work room of the trailer where I was cataloguing fossils from the day's collecting. There was a look of concern on his face. I tried to guess what the problem might be before he started talking. I had noticed some tension between the Germans and the Americans. For example, when a fragmentary specimen had been found one day by the German contingent, one of the group had simply sunk a shovel into the ground, scooped it up, and handed the pieces to the technician. Some of the Americans had been horrified because the specimen had not been carefully hardened, supported by a plaster jacket much the same way an orthopedist sets a broken bone, and then carefully removed. A discussion had ensued. I thought that everyone had then reached an understanding and that the ruffled feathers had been smoothed, but you could never be sure. I braced myself for another bout of international scientific diplomacy. But that was not the problem. Instead it was morale. The German contingent was beginning to feel that we were approaching the point of diminishing returns. They thought that the fossils we were finding were too fragmentary and of dubious significance.

This opinion caught me unawares. I had to admit, however, that I, too, had been somewhat surprised that so many specimens were fragmentary after my earlier hypothesis regarding the low-energy depositional conditions. But if Africa teaches one anything it is patience. Louis Leakey had said the same thing and had toiled under the African sun for over twenty-five years before the complete skull of *Zinjanthropus* had turned up in 1959. We, however, had only been in the field for a few weeks, and I told Wahid that it was too soon to give up. But we decided to have a meeting and talk about the issue.

We had the meeting in the workroom, where I spread out on the table the specimens, now all nicely numbered, glued together, and concentrated in one place. I arranged them by animal groups, some of which had not been found before at Sahabi. In this room of the trailer, a space of human proportions, what we had collected did look impressive. It was all too easy to lose one's perspective in the immensity of the desert. Single sand-blasted bone fragments in the desert were pretty uninspiring. This change in perspective seemed to help. Another point that came out was that the Germans for the past several years had been working primarily on invertebrate fossils, such as ammonites, that, when found, are frequently found in abundance. Fieldworkers then spend their time finding the best and least damaged museum piece. After we had talked, everybody seemed to feel better about the project. The meeting was a watershed in German–American relations. For the rest of the field season there was a much closer working relationship between the two contingents. We decided that there would not be any more single-nationality surveying forays.

Although the meeting had served to create a certain esprit de corps, I was under no illusion about the conditions at Sahabi: they were difficult, very difficult. The heat, especially during the middle of the day, was almost unbearable. The sand was everywhere and in everything. The whining of the wind was incessant. The flies came from nowhere to attack your eyes, nose, and mouth, attracted by any hint of moisture. And the amenities at Sahabi were meager. We did have hot water for showers, but hot was the only temperature available and there wasn't very much water. The food was almost all from cans and was prepared simply because we could find no cooks to hire. Libya had a labor shortage and used mostly foreign nationals for the construction and service sectors of their economy. In later years we were able to hire Pakistani and Chadian cooks and excavation workers. Libya, of course, officially did not allow alcohol of any kind so there was no chilled Chablis for dinner. Any one of these hardships or the combination could bring you face-to-face with yourself in the desert and make you question why you were there.

We all had our own pet peeves, but the frustration was more than that. It was a particular resonance of personal susceptibilities with environmental circumstances that could lead one to the breaking point. Some personal susceptibilities obviously would have been a major problem, such as fear of open places—agoraphobia. An agoraphobic would not even have made it to Sahabi, let alone been able to work there. All our expedition members were quite normal, but as time, the environment, and fatigue took their tolls personal foibles became more obvious. With Ray one day, the last straw had been flies. While sitting at the worktable, he suddenly stood up, took off both his flip-flops, and started madly swatting at all the flies in the room. He succeeded in smashing a few, but most buzzed around calmly until he sat down again. With Lierl, the German technician, the greatest trial had been beer, or the lack of it. Walking by his tent one day, I glanced through the open tent flap to say hello. He wasn't there. The tent was spartan, but tacked onto the central pole was a torn-out page of a magazine ad, striking in its surrealistic simplicity. It was a single ice-cold bottle of Heineken. With a graduate student on a later expedition, the greatest ordeal was lack of cold water. Because he refused to drink the warm water in camp and was running the risk of serious dehydration, I had no option but to send him back to Benghazi where Wahid put him up in his house until the student felt that he could return to the desert.

Many of the new species we discovered at Sahabi in the first and subsequent years were small animals. The earlier paleontologists who had worked at Sahabi had had the usual biases of their generation. They had searched for and found only the largest animals. Medium-sized animals such as hominids would have eluded them. Small animals such as rodents were so small that they never would have been seen. We looked for all sizes of bones and teeth, passing several times over a potential fossil site, each time with a different search image in mind. And the very smallest animals were collected in fine-mesh screens from sediment washed from potentially productive sites.

Using these techniques we found the first primates from Sahabi in 1978. The first specimen was a monkey half jaw with teeth found at locality P28. It was an exciting day. Wahid, Salah, and I were surveying near Elephant Hill, where two tusks of *Stegotetrabelodon* earlier had been found sticking straight out of a hill. Wahid and I had found a tusk of a totally new type of proboscidean first thing that morning. It was a "shovel-tusker," an elephantine animal with no upper tusks but dual, flat, scoop-like lower tusks. It had never been found at Sahabi before. While we were engaged in collecting all the surface fragments of this specimen and wrapping them up, Salah went across to look around in another area I had earlier named P28. Just as Wahid and I were preparing to find Salah, he showed up. He had something wrapped up in toilet paper that he first handed to Wahid, who then handed it to me. It was a monkey, our first primate. I was elated. It was another first for Sahabi and another significant find, because little was known about monkeys in North Africa during this time period. The question of their evolution was intriguing in itself, but the new fossil made my thoughts turn to their primate cousins, the apes and the hominids. Would we find any?

The 1978 field season drew to a close, and although the faunal inventory had grown impressively we had no hominids. Other paleontologists find this monomaniacal obsession with fossils of one specific lineage, our own, rather odd. An entire, beautifully preserved fauna was being exposed to our view, but this anthropological preoccupation with hominids tarnished the appreciation of it. Daryl Domning, our expert on sea cows, on the other hand, was in seventh heaven. Entire fields of partially articulated sea cow skeletons still lay where they had died at Sahabi, having been stranded on tidal flats millions of years ago. Wahid could not have been happier. Proboscideans were all over—stegotetrabelodons, mastodons, gomphotheres.

Wahid could not understand why I would not switch my interest to another group, say micromammals. Why try to study something that isn't there? Wahid had thought about this suggestion. What, he thought, drives Noel to keep going out in the desert every day? Maybe, he concluded, it is the difficulty of the search. Micromammals certainly are hard to find. Munthe had to work hard washing all those tons of sediment and then peering through a microscope to find those tiny specimens. They weren't impressive animals like elephants, but then neither are primates. Wahid had a point, and I tried to explain it, as much to clarify my determination in my own mind as to give him some reason for what must appear to be a rather fruitless exercise. Scientists had been trying to figure out the animal origins of the human species since Darwin, and the answer to a question of this magnitude might take decades or a lifetime to discover. You could say something like this to Wahid and he would understand it. That's why I liked him. He shrugged his shoulders resignedly and accepted this explanation as one might accept a friend's unrequited quest after an unattainable member of the opposite sex.

Even without hominids, Sahabi was important anthropologically. Since the 1960s, a major emphasis in paleoanthropology had been understanding the environments of early hominids. It was important to reconstruct what sort of climate, vegetation, landscape, water resources, and animals were around so that some interpretation of the adaptation and behavior of the hominids could be reconstructed. For example, the evolutionary arguments on the origin and function of walking on two legs, bipedalism, are very different depending on whether the adaptation first developed in a forest versus a savanna. Sahabi was providing abundant paleoecological data for a time period and part of Africa that had been very poorly known, and for that reason I felt that the project was very worthwhile. I also felt confident that hominids would be found. To have monkeys, pigs, elephants, antelopes, hippos, and all the other animals that co-occurred with hominids in the later East African sites, but no hominids, seemed unlikely to me.

The 1978 field season ended, and everyone looked forward to going back to whatever they had longed for in the desert— Lierl to his beer, and each to his own. I prided myself on having none of these strong aversions to any of the conditions, normal or abnormal, that might come up in the course of professional work. I also was physically strong and almost never got sick. A camel stew that had put everyone else in bed had

The Cold Spring Harbor conference in 1950 marked a turning point in anthropology, as a group of the "Old Guard" met up against the "Young Turks" who wanted to supplement the traditional methods of fossil identification, measurement, and nomenclature with a bewildering variety of methods including ecology, ethology, and population biology. Above, the ever dapper Earnest A. Hooton, professor of anthropology at Harvard and mainstay of the Old Guard, slips the intellectual jabs of Joseph B. Birdsell, who along with Sherwood Washburn led the Young Turks into the fray. The "new physical anthropology" that emerged saw its first full application in paleoanthropology at the site of Omo in southern Ethiopia (below), where Louis Leakey and F. Clark Howell initiated a multidisciplinary project which offered rich evidence for the new scientists.

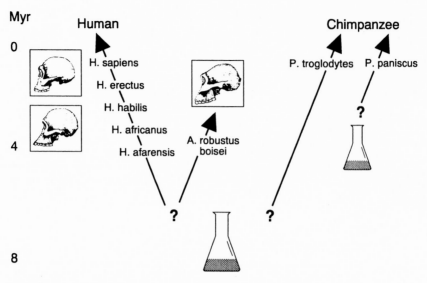

After six decades of discovery, interpretation, and reinterpretation of the fossil record, a widely accepted consensus has been reached of species evolution along the family branch specific to humans (above, left). But when and how did this branch diverge from its relative, that of the two modern chimp species (above, right)? Only the new methods of molecular research, represented by the chemical flasks, provide answers for the evolution of the chimp, whose ancestry remains totally undocumented in the fossil record. The author's hypothesis (below) combines climatic evidence with the dates and anatomical studies of hominid fossils to argue that the formation of the western African Rift Valley, 7−9 million years ago, triggered human evolution to the east, in the drier savanna, while chimps evolved to the west, in the forests and woodlands.

## Rifting in Africa and the Chimpanzee
## Split at 7 – 9 ma

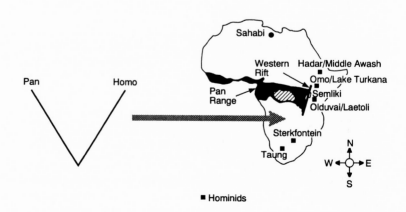

The author's expedition to the Sahara Desert site of Sahabi in Libya (left) provides our best glimpse to date of northern Africa following the drying up of the Mediterranean Sea between 6.2 and 5.0 million years ago. Despite the modern desertic conditions, Sahabi at 5 million years ago was a savanna fed by a perennial river surrounded by forest. The environment teemed with animals, much as the modern Serengeti Plains of East Africa, but with an abundance of species which to our eyes would seem strange.

Very early hominids may have been a part of this scene but so far the three hominoid specimens discovered have been too fragmentary to draw firm conclusions. Below is the infamous hominoid collar-bone (or dolphin rib?) discovered by the author, placed for comparison under a modern human collar-bone.

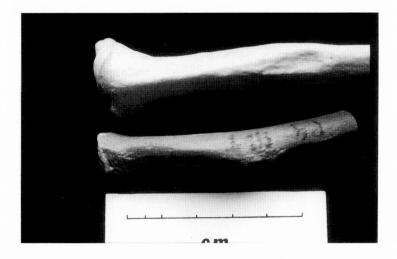

The Western African Rift Valley running between Zaire and Uganda provides one of the most important evolutionary laboratories for testing hypotheses of human evolution. In the Upper Semliki Valley of eastern Zaire the author and a long-term colleague, geologist Jean de Heinzelin, investigate the fossil site of Senga 13B. This site showed that early hominid environments in Central Africa had been open savannas with nearby forest at 2.0 – 2.3 million years ago.

As research into earlier deposits in the Western Rift continues, we expect to encounter the ancient forest habitats of the hominoid populations from which sprang not only the line leading to the chimpanzee but our own lineage, the hominids. The author's current major project, in Kenya (below), offers the potential of hominid discoveries that could help fill huge gaps in our knowledge. But the only certainty is that our hypotheses will continue to evolve as new data is discovered.

left me unscathed. But as my great-grandmother would have said, "pride goeth before a fall."

––––––––––

My nemesis lay not in the wastelands of Africa but in the wastelands of New York City. When the 1978 Sahabi field season was over, I did not go back to UCLA but to New York. I had accepted a job at New York University offered to me by the chairman of the Anthropology Department, John Buettner-Janusch. I had known Buettner-Janusch since I had been an undergraduate, when he had given me advice about graduate school. He had then been at Duke University, where he had founded and directed the world-renowned Duke Primate Center. NYU had recruited him to put the Anthropology Department on the map, and he wanted me to bring in an active, high-visibility paleoanthropological research program. The International Sahabi Research Project based at NYU was exactly what he had in mind. Buettner-Janusch had made me an attractive offer: a new lab to be built to my specifications, a joint appointment with the Anatomy Department in the Medical School, a highly sought-after apartment with parquet-wood floors just off Washington Square, and a salary several thousand higher than UCLA. And the American Museum of Natural History and a number of good colleagues were in the area. At UCLA I faced an uphill battle for space and resources in a department dominated by social anthropologists, hostile to developing a strong program in paleoanthropology. The NYU offer was clearly better, but I had reservations. It was Buettner-Janusch who convinced me. An unabashed New York chauvinist, he had answers, solutions, or a good-humored shrug of resignation for every one of my negative points about the Big Apple. I accepted NYU's offer. It was a decision that I was to regret, but not until after the Sahabi project had completed its field program several years hence.

Returning from Sahabi, we busied ourselves with getting the laboratory aspects of the Sahabi work organized and under way. Construction started on the new lab, but in the meantime I was stuck in a closet-sized office looking out into a dark air shaft. The room somehow reminded me of a Gestapo interrogation room. To counteract this feeling Dody, my former wife, and I covered two large vertical pipes in the room with brown crepe paper and made cutout palm fronds that I glued to the top. Such an act of frivolity was unusual for me. I should have recognized the similarity with Lierl's picture of a Heineken bottle in the Sahara, but didn't. I was to work at NYU for another five years.

## THE ANALYSIS OF SAHABI

The importance of the Sahabi discoveries began to become more obvious to us as lab analysis proceeded and the various specialists reported back on their conclusions. Sahabi showed a unique mixture of animals. We had discovered a new world in time and space. The most abundant animal was the anthracothere, a creature like a hippo but not closely related, named *Merycopotamus*. An Italian geologist who obviously knew nothing about vertebrate paleontology had found a skull in a dry gulch north of Sahabi known as the Wadi Faregh just before World War II. He thought he had found a dinosaur, which he named *Libycosaurus*. Later the name was synonomized with the anthracothere *Merycopotamus*, also known from earlier beds in Tunisia. Anthracotheres had been extinct in the rest of Africa for over 10 million years. That they had held on and even flourished in North Africa was testimony to the unique zoogeography of the region.

In 1978 we found the first bear from Sahabi, which turned out also to be the first bear anywhere in Africa. Another record of bears came from a site in South Africa, also near the sea, known as Langebaanweg and dated to about 4 million years ago. At some point later, bears died out in Africa and are not found anywhere on the continent today. The other carnivores were abundant and diverse. There were four different species of hyenas and several large cats. I thought that the prevalence of carnivores might be a reason that most of the bones were so fragmentary. Perhaps the hyenas and other large carnivores had been so abundant that any dead animals lying on the ground were chewed up pretty completely before they could get into a fossil deposit. Certainly quite a few bite and chew marks were left on the bones. The smaller animals, like primates, would have suffered the most from this treatment.

After the success of the first field season, interest in Sahabi soared. We published papers on the first year's findings in *Nature* and in *National Geographic Research Reports*, the Rolex Awards for Enterprise recognized the project and put it in their book *The Spirit of Enterprise*, and Libyan TV put Wahid and our lab in Benghazi on the air. Our grant to the National Science Foundation to return to Sahabi was approved.

We could have done without some of the interest. Senator William Proxmire contacted me by letter to request a long list of explanations of why the American taxpayer should support our research in Libya. I was being considered for a Golden Fleece Award! Proxmire and his staff had taken it upon themselves to ferret out wastefulness in the federal government. Among their former awardees were the Pentagon for buying extremely expensive toilet seats. We did not know if we were being considered because of tense U.S.–Libyan relations or because of the nature of the research. In the research arena we knew, as did most in academia, that Proxmire was particularly critical of social science research. The Anthropology Program at the National Science Foundation, which was funding our research, awarded grants that ran the gamut from social science, such as the ethnography of graffiti on public bathroom walls, to natural science, such as excavation of 40-million-year-old primates. Because we were far over on the natural science side of the spectrum, I thought we were safe. I gave Proxmire's letter to my graduate students and asked them to formulate a response as a lesson in communicating what we did to the public. I thought they did a very good job, and apparently so did the senator because we never received the Golden Fleece.

The 1979 field season brought the collection of many new fossils and significant advances on the geological front. But at the end of the field season no hominids had been found. There were only a few specimens in the "to identify further" category. These were specimens that were too strange or too rarely found to identify on sight and that had to be taken back to the museum for comparative study. A bone from locality P4 picked up by the geological team was to become the most infamous of any fossil found at Sahabi. I entered it in the field register as "?rib or limb element."

When we were back in New York I took the box of "to identify further" bones with me to the American Museum and began the process of sorting out what they were. Most of them did not take too much time. Within a few hours I had determined what they were or most likely were: carnivore phalanx, bovid frontal bone fragment, anthracothere vertebra, bird femur fragment, pig unciform, and so forth. But the P4 bone defied categorization. It was too curved and asymmetrical to be a finger or toe bone, even of a digging animal like an aardvark. It looked like no rib of any mammal or reptile in the collections, and it was certainly much too large to be a bird. I even tried comparing bacula, the penis bones that most mammals have, but the Sahabi specimen was the wrong shape and again was too curved. It looked something like a clavicle, the collar bone, but I gave up after several hours and put the bone into the box marked "indet." for "indeterminate." It would have stayed there if its surface had not been so well preserved and if there had not been so much anatomical detail. It was in good enough condition to identify, and I kept pushing to figure out what the P4 specimen was.

The next several times I went to the museum I looked at all the sea mammals, particularly their limb bones and ribs, to see if I could find elements that might match. And I started looking at clavicles. Not too many animals have clavicles. Running animals like horses, bovids (antelopes), and pigs have lost theirs through evolution. Only animals that exert strong side-to-side movement of their arms, like diggers or climbers, have well-developed clavicles. The Sahabi specimen was too large to be any of several species of digging mammals and too large to be a monkey. If the specimen was a clavicle, I was drawn to the conclusion, although tentatively, that this specimen represented a hominoid, that group of large primates that included hominids and apes. At first I rejected the idea because I knew that the first identification of an ancient hominid or closely related form from North Africa based only on a collar bone, and not the more traditional jaw, skull, or tooth, would provoke debate.

For several months I took the specimen to other museums to make wider comparisons. I went to the Smithsonian where I again went to look at the sea mammals, and I asked the curator who studied seals what element he thought the bone was. He kept it for an hour or so and then suggested that he thought it was closest to a human clavicle. I took the specimen to Berkeley and discussed its identification with a number of colleagues who were there visiting for a meeting. A clavicle seemed like the best guess, but the bone was missing the outer third of its shaft, which could have confirmed the identification.

## The Flipperpithecus Affair

I decided to submit the P4 bone to every possible comparative test with the entire range of variation of hominoid clavicles to try to resolve the issue. If I could make a convincing case for the specimen's hominoid status, then I would feel justified in taking the unorthodox step of publishing a collar bone as our first record of a hominoid. I sectioned a human clavicle at the point that the Sahabi bone was broken and compared curvature and inside bone structure. I did an exhaustive literature search that gave me the ranges of variability in human and primate clavicles. I found an explanation for every anatomical characteristic on the Sahabi specimen in terms of the specimen representing a clavicle. This was the crucial decision. If P4 was a clavicle, then the only animal it could represent was a large primate, probably a hominoid of some sort. I weighed the advantages of waiting to publish the P4 specimen as a possible hominoid until we had more fossils that could be ascribed to the same general category. I finally decided that the most scientifically straightforward course was to publish all my observations and state my conclusions in the most conservative and supportable way that I could. I realized that others had been mistaken about identifications of clavicles. Louis Leakey and his team, for example, had misidentified a turtle humerus as a hominoid clavicle in the 1940s. But after a great amount of effort and consultation, I thought the most reasonable identification for this bone was a hominoid, and publication would make this known. If someone had additional information and a different opinion, then they could publish that opinion. This dialogue, after all, is what scientific debate is all about. I submitted the paper, and it was published in the *American Journal of Physical Anthropology* under the title "A Hominoid Clavicle from the Mio-Pliocene of Sahabi, Libya."

Tim White decided to respond to the paper. White was still smarting from the controversy over Lucy, in which I and my colleagues had successfully argued that White and his co-author Johanson were incorrect in dating Lucy so far back as they had (see Chapter 4). White had been defeated in that debate, but he now saw a chance for a counterattack. One of White's graduate students who was preparing for his Ph.D. oral exams asked for a cast of the Sahabi P4 specimen through an old friend of mine at Berkeley. My lab had just finished making a number of casts, and it was sent out. We would have sent the cast directly to White had he requested it, but he apparently thought the surreptitious approach was preferable.

Several months later I received a manuscript to review from the *American Journal of Physical Anthropology*. White was the senior author. Its point was simple. P4 was not a clavicle and therefore not a primate. It was instead a rib, specifically a dolphin rib. I was not surprised by this conclusion because I already had seriously entertained this possibility and had ruled it out. A number of the blanket statements about ribs and clavicles that White et al. made could not be supported by references to the literature, but my real quarrel was with the tone of the paper. White made reference to the fraudulent Piltdown skull, obliquely implying that I had intended some hoax or coverup. I was infuriated. But, controlling myself, I wrote back to the journal saying that the paper should be published if this unprofessional slur were removed and if there were fuller referencing of the authors' observations. I also wrote to White and gave him an accounting of the points about the specimen on which I directly disagreed with him. To counter the implied claim that I had not been fully forthcoming with all the information on the Sahabi identification, I presented a poster session at the next professional association meeting with all the comparative photos and measurements supporting my arguments. White never responded to the scientific arguments.

When White's paper came out, the press had a field day. *Science News* proclaimed, "Hominoids Become Porpoiseful." Reporting on a conference in Berkeley in which White showed slides of Flipper the Porpoise and sarcastically introduced the new name *Flipperpithecus boazi*, the *San Francisco Chronicle* ran pictures of me and White. National Public Radio covered the story. There never was any real debate. White simply maintained that the bone was not a hominoid clavicle and was a dolphin rib, because he knew it was. He would then launch into an ad hominem attack to whomever would listen. Don Johanson retold the incident in his book *Lucy's Child* and reported that White responded by noting that his attacks were actually "ad delphinum" (against the dolphin).

The Flipperpithecus affair was an unfortunate tarnishing of the image of the Sahabi Project, but I felt that I had done what I should have in publishing my observations. I had to agree with the general consensus that further remains had to be found before the presence of fossil hominoids at Sahabi could be accepted.

## BACK TO THE DESERT FOR THE ANSWERS

Meanwhile we continued the field research program at Sahabi. Three more paleontological field seasons and several geological excursions were carried out until 1981. Our last field season, although we did not know it at the time, was the winter of 1980–1981. We were under pressure to discover some more definitive evidence bearing on the hominoid controversy. I wanted to cover the maximum amount of territory in surveying. Our grant was renewed, and I used the $6,000 budgeted for vehicles to buy two used dune buggies. I thought that their lightness would be an asset in driving over the loose sands and sebkha deposits where the Landrover occasionally bogged down. The dune buggies were rebuilt Volkswagen beetles with fiberglas bodies. We bought them from a dealer in North Carolina, had them shipped to New York, and put them on a boat to Benghazi.

Our 1980–1981 field season was during the winter. After dealing with the heat, sandstorms, and *ghiblis* (the superheated winds also known as siroccos) during the summer, I was looking forward to less extreme and even cool working conditions. But winter in the northern Sahara is a wild time. The sandstorms were even stronger and more numerous, flattening the tents at least once a week no matter how we tried to shelter and secure them. The storms were at times so severe that we were forced to spend whole days inside the trailer, unable to see outside for more than a few feet. It was not cool. It was usually cold. Several times it snowed lightly and hailed.

Our last field season was an entirely male expedition that lasted for three months. I had received a Presidential Fellowship from New York University, which had given me a semester off from teaching, and my fellow expedition members also had been able to get release time. Daryl Domning and Doug Cramer both taught in medical schools and their anatomy courses were over, Jens Munthe was in between a teaching job and a new position as an oil company geologist, and John Page from Westminster College in Missouri had received a special National Science Foundation grant for small college professors to gain research experience on large NSF-funded projects. Wahid and Ali came down to Sahabi occasionally in between their classes and administrative responsibilities in Benghazi, and Jean de Heinzelin was to join us later in the season. The team was a no-nonsense group with years of field experience, except for Page who had never been to Africa before.

We made a strange-looking group leaving for the desert early in the morning. Cramer led the expedition down the road. A large muscular man with a full prematurely gray beard, he had bought a traditional long black Libyan hooded cape, which he wore. People by the roadside were first alerted to our approach by the sound of Cramer's vehicle, a roaring green-spangled

dune buggy with chrome pipes and a souped-up engine modified by good old boys in North Carolina. When they saw that it
was being driven by a huge bearded man in a black cape, their
mouths dropped open. Page, who was driving the second, red-
spangled dune buggy, had to swerve more than once to avoid
hitting pedestrians who had stumbled into the road to stare
after Cramer. I brought up the rear in the Landrover.

Humor proved an important element in survival, and
Cramer and Page became our comedy team. Page was the
straight man and Cramer delivered the punch lines. But perhaps their funniest routine was an entirely unplanned escapade
that happened on a public holiday, Mohammed's birthday,
when we went into town for supplies. As Wahid and I went into
one of the few stores that we found open, Cramer, who had
just been complaining about not enough public toilets in Libya,
went off to pee. Page decided he needed to go, too. As Wahid
and I were emerging from the store, we saw an angry mob pursuing Page and Cramer. My first response was an attempt at
conciliation, but Wahid quickly sized up the situation and said
to get into the Landrover and get out of town. Page and
Cramer jumped into the car and we sped off, leaving behind a
group still running after us, shaking their fists and throwing
rocks. Cramer, shaking his head, said that these people must be
awfully picky about public sanitation. Then Wahid explained.
Apparently Cramer and Page had chosen as their pissoir a construction site that had just been consecrated for a new mosque.
The fact that it was one of the holiest days of the Islamic year,
Mohammed's birthday, had further inflamed the crowd. When
we had begun to recover our composure after a paroxysm of
laughter, Page said that he had tried to apologize, but nobody
had understood English.

The field season was our most productive. We discovered
many new localities as we covered new territory and resurveyed
old territory that now had a new topography thanks to the shifting of the sands by the winds. I began walking more and even
jogged long distances over the desert because many of the localities could not be seen until you were almost on top of them.
They were distinguished by only the slightest rise in topography

and perhaps the slightest hint of rock beneath the sand. I found P99 as I jogged back to camp from a broken-down dune buggy. We had driven within a kilometer of the area hundreds of times. In one day I found what turned out to be new records for species of pig and monkey at the site. Both fossils were well-preserved jaws amidst many other fossils. P99 is where a complete hominoid cranium may be found at Sahabi, but the few days remaining in the 1981 field season did not allow us to fully survey the locality and to put in an excavation.

Two specimens, however, came out of Sahabi that threw light on the hominoid controversy. The first was a portion of skull that was too round and flat to have been either of the two monkeys, *Macaca* or *Libypithecus*, that we had identified from the site. Instead it looked most like a gibbon skull. Gibbons are the smallest of the apes and have a relatively larger brain/body size ratio than monkeys. They are today confined to the forests of Southeast Asia. Of course, the Sahabi specimen probably was not a gibbon, but its rounded skull vault implies a relatively large brain, and adding in its small size, there is a gross similarity. We would have to have the creature's face, however, before we could tell what it was.

The other specimen was a piece of the lower outside leg bone, the fibula. After an exhaustive comparative study of fibulae, similar to my study of clavicles, I concluded that this specimen looked surprisingly like humans and very unlike modern apes. If correct, this conclusion implies that the creature, though small, had an ankle joint like humans and might have been able to walk on two legs. I published these two specimens along with an extended discussion of the clavicle in the monograph that we compiled on the Sahabi geology, flora, and fauna.

I felt that the discovery of the fragmentary skull and fibula from Sahabi had opened a new chapter on the hominoid debate in North Africa. But White apparently had lost interest in the issue and made no response other than a rather sulky review of the Sahabi monograph. He had now turned to investigations of interpersonal aggression in early humans. Sahabi received only the briefest mention in one of the leading textbooks and in a review of early hominid sites in Africa by pale-ontologist Kay Behrensmeyer. In an article published in 1988, paleoanthropologists Andrew Hill and Steve Ward left the hominoid nature of the Sahabi specimens as open questions. The scientific community was not ready to accept the Sahabi fossils as hominoids until more complete remains were discovered.

The political situation between Libya and the United States was worsening by the time I left Libya during the last field season. I went back to Libya once more to complete the writing of the monograph with Ali and Wahid, but I had to leave when President Reagan issued a proclamation that all Americans should leave Libya for their own safety. I knew perfectly well, however, that there was no danger to Americans on the street in Libya, as this proclamation implied. As soon as I had arrived back in New York I wrote a letter to *The New York Times* protesting this breaking of a beneficial cultural and scientific tie between the United States and Libya. Shortly thereafter, Reagan ordered U.S. planes to bomb Tripoli and Benghazi. All of our pending grant proposals to return to Sahabi for more excavation were immediately suspended.

## METHODOLOGY AND SAHABI

The real importance of Sahabi has lain not in what few fossils of primates have come from the site so far but in what the investigation says about the way hypotheses are framed and how research is undertaken in modern paleoanthropology. The old research methodology might be called the Indiana Jones School of Paleoanthropology. This approach focuses entirely on the hominid fossils and puts the geology, paleoecological studies, flora, and fauna on the back burner. When Indiana Jones successfully negotiates the booby-trapped cave and secures the crystal skull in *Raiders of the Lost Ark* his expression of fulfillment indicates that he has attained his objective. All the information that is relevant to the object is inherent in the object. Context, in this case the temple in the cave, does not matter.

Most paleoanthropologists today belong to what may be termed the Contextual School, in which the context is viewed as centrally important to formulating and testing hypotheses of hominid evolution and behavior. This approach has led to graduate students completing theses and dissertations on such diverse groups as horses (as Ray Bernor did at UCLA) or hippopotamuses (as Paris Pavlakis did at NYU) in anthropology departments, paradoxically concerned with humankind. The evolutionary history of these groups of animals can reveal much about the environment, age, and ecology of the habitats they shared with hominids. The scientifically detailed context that Sahabi provided for the North African Pliocene of 5 million years ago (not 7 million as we first thought) was our first real glimpse into an area of a continent and a time period that we know were important to the earliest emergence of our lineage.

Beginning in the 1970s when the Messinian Event was discovered by Deep Sea Drilling Project excursions in the Mediterranean, there has been speculation about how such a major change in the circum-Mediterranean world might have affected hominid evolution. A cartoon circulated by geologist John Van Couvering showed a knuckle-walking ape venturing out onto the hot Mediterranean salt flat, hopping up on two legs and shaking his fingers to cool them off. "The origin of bipedalism," the cartoon read.

More serious suggestions were made as well. For example, C. K. Brain of the Transvaal Museum suggested in 1980 that the Messinian Event and the drying up of the Mediterranean would have set into motion a number of climatic changes in Africa that resulted in the spread of savannas. This environmental change was, according to many scholars, one of the most important in initiating the evolution of the bipedally walking hominids from forest-living African apes. And the timing of the change accorded well with the postulated dates of the splitting of hominids and African apes derived from the new molecular studies. Sahabi was immediately adjacent to the Mediterranean basin and dated to the end of the Messinian Event. It thus is positioned perfectly to test whether this environmental change occurred and what evolutionary effects it may have had.

Another evolutionist, Elisabeth Vrba, has suggested that faunas should change radically when the environment changes radically. This hypothesis is an ecological variant of the punctuated equilibrium model of evolutionary change of Stephen Jay Gould and Niles Eldredge. Vrba suggested that the boundary between the Miocene and the Pliocene Epochs was just such a period of punctuational change and that the first hominids were part of the evolutionary response of the fauna. Again Sahabi seems like the perfect place to test the model.

The fauna from Sahabi did not support the punctuational model. The carnivores seemed like those known from the earlier Miocene of western Asia, and the anthracotheres were left over from much more ancient times in Africa. They had disappeared entirely from sub-Saharan Africa 12 million years before but had held on, evolving slowly on their own track in North Africa. The proboscideans, too, recalled earlier faunas, with the four-tusked *Stegotetrabelodon* and the primitive shovel-tusker *Platybelodon* present. I published a short paper that rejected the punctuational model for North Africa and Sahabi in the Pliocene.

Perhaps that assertion was a bit like throwing out the baby with the bath water. There was an alternative explanation. Sahabi had a perennial river that flowed into the Gulf of Sidra, then probably a hypersaline pool. This river was surrounded by abundant trees that were preserved as drifted logs and identified in thin-section by our paleobotanist Roger Dechamps. What if the river provided a refuge for the older fauna, species of which were able to hang on in a part of the environment that had remained unchanged, while away from the river it was a brave new world of savannas and new species? When the river disappeared because something had cut off its source waters far to the south so did the water-dependent old fauna, leaving the future to the new species. Hominids could well be part of this new wave of evolution. Answers lie in the shifting sands of Libyan–American relations and hidden among the dunes at Sahabi.

The paleoanthropological results from Sahabi have been far from conclusive, but they have made me question much of the received wisdom about the origin of the hominids. If ancestral hominids were small-bodied and if they were already bipedal at 5 million years ago, as the Sahabi fossils might imply, what would these assumptions do to conceptions of hominid origins? Standing and walking upright—bipedalism—played a big role in all the scenarios of the original split of the hominids. What are the implications if the date of bipedalism is pushed back in time and the split of the apes away from the hominid line continues through molecular studies to be pushed forward in time? We investigate this and related questions in the next chapter.

# 6

# Origins of the Featherless Biped

The ancient Greeks sought to distill a succinct but accurate definition of humanity out of the panoply of nature. Aristotle chose that characteristic that sets people apart from almost all of the other common creatures on the planet: the fact that humans walk upright on their hind legs. But it was one of Aristotle's students who pointed out that the definition of bipedalism as the defining trait for humanity does not distinguish humans from that class of hopping, walking, and flying animals known as birds. The Greeks amended the definition to "featherless biped" to uniquely distinguish humanity from all the animals known to them in antiquity. Today, of course, we know of other featherless bipeds, such as kangaroos and the dinosaur *Tyrannosaurus rex*, but the basic definition still has heuristic value.

## BIPEDAL APOLOGIST AND BIPEDAL ADVANTAGIST SCHOOLS

Bipedalism is still the defining characteristic of the Family Hominidae. All theories of human origins have had to postulate how and when the hominoid ancestors of humans stood upright. Darwin believed that bipedalism evolved in ground-living humans because it freed the hands for tool use, including the use of weapons for defense. Many other ideas have since been proposed. Among these hypotheses are:

- By standing upright, hominids could see over tall grass, thus avoiding predators and finding prey on the savanna grasslands.
- Because the upright stance is used in all hominoids for aggressive threat, by permanently standing up hominids ensured themselves a place in the hurly-burly of the crowded carnivore ecology of the African savanna.
- Hominids became bipeds because they needed to carry things—helpless young infants or food—for relatively long distances.

All these hypotheses have something in common. They all ignore the most obvious possibility that bipedalism evolved as an efficient locomotor adaptation for moving hominids around in their environment. Since Darwin's time bipedalism has been considered a very inefficient mode of locomotion. You need only think of trying to catch your cat, trying to keep up with your dog running after a ball, or even vying to catch a greased pig at a country fair. Quadrupeds are quite superior to humans when it comes to speed, quickness, and maneuverability. I term this point of view the "bipedal apologist" school of thought. The bipedal apologist theoreticians feel that bipedalism is such a slow, ponderous, and generally poor locomotor adaptation that special selection for other attributes associated with it must be found in order to account for its evolution.

I had generally bought the bipedal apologist line of argument until I heard a paper at the annual meeting of the American Association of Physical Anthropologists over ten years ago. The paper was by Donald Mitchell of the State University of New York at Buffalo, a physical anthropologist and marathon runner. Mitchell had helped to organize the local marathon race and had computerized the start and finish times of all the runners—male and female, all different ages, all different weights. His results were surprising. He showed that even the slowest human runners could, with even a slight head start, outrun lions, cheetahs, leopards, hyenas, and wild dogs, not by speed but by out-distancing them. Here, then, was a positive selective advantage for bipedalism on the African savannas. Maybe hominids adopted bipedalism because it was an efficient long-distance mode of locomotion, not fast but reliable—the biological version of the Volkswagen beetle. Comparative physiological tests have now shown convincingly that walking humans use less energy to traverse a distance over a flat surface than do most quadrupeds. We might term the theoretical position arising from these conclusions the "bipedal advantagist" school. Paraphrased simply, there was a strong positive reason for bipedalism to have evolved sui generis in the earliest hominids.

What is even more germane to our concern is that human walking is vastly more efficient for moving over flat terrain than is chimp or gorilla-style locomotion, a type of quadrupedalism known as knuckle-walking. The other apes do not do any better when moving along the ground. The orangutan fist-walks, supporting the weight of its upper body with a fully closed hand on the ground instead of with the bent fingers of the chimp and gorilla. The gibbon, the lightest of the apes, totters along on two legs with its elongated arms raised akimbo, on the rare occasions when it comes to the ground. All of the ways that apes employ in moving on the ground are costly in terms of energy. They therefore cannot be kept up for long.

Hominids evolved their unique mode of locomotion to allow them to move long distances along open terrain. This evolutionary innovation likely was caused by a climate change from wet to dry in the area of Africa where the hominids originated. Changing vegetation patterns forced the forest-living ancestors of hominids to traverse open stretches of savanna to reach other life-sustaining patches of forest. Those populations able to successfully make the traverse survived. Those who were caught in shrinking forests with dwindling food reserves died out. As the climate continued to get drier, savanna increased at the expense of forest and the distance between dense patches of trees became greater and greater. Mortality increased in the ancestral hominids until only the few that had the best means of making it over the open terrain survived. This is the way that natural selection works, and bipedalism became the best means of locomotion in the hominids.

If this scenario is correct, then hominids must have arisen from a primate that was adapted to arboreal locomotion of some sort. Was this a form of ape locomotion, some type of climbing, hanging, and swinging from branches of trees, which is referred to as brachiation? Or was bipedalism derived from a more primitive, monkey-like form of locomotion: running along branches on all fours, a type of locomotion known as arboreal quadrupedalism?

## THE ANTECEDENTS OF BIPEDALISM

In the absence of fossil data, anthropologists have used anatomical and behavioral clues from living primates to speculate on the antecedents of bipedalism. Sherwood Washburn is a proponent of the knuckle-walking hypothesis for hominid bipedalism. He notes that speakers at lecterns habitually adopt a knuckle-walking posture as they support their weight with their hands and that football linemen are perfect examples of gorilla-like knuckle-walkers. Another of Washburn's examples always reminds me of the schoolyard prank in which you say, "One sign of stupidity is hair on the knuckles." As your victim then looks at his knuckles, you continue, "The other is looking for it." Washburn also pointed out that there is no hair on the skin over the second phalanx of the human fingers. "Just like chimps," he would say.

But there is also support for the position that hominids did not pass through a knuckle-walking phase in their evolution. The dean of American physical anthropologists, Ales Hrdlicka of the Smithsonian, was a firm believer in the quadrupedal origin of hominid bipedalism. He published a detailed paper on the unlikely topic of "Children That Run on All Fours" to demonstrate that this form of human locomotion is still quite efficient and well within the behavioral repertoire of modern humans. He collected anecdotal evidence from parents whose children had developed this form of locomotion and noted the speed with which they could move. Frequently they easily could outrun their parents. Another supporter of the quadrupedal origin argument was William Straus of Johns Hopkins University, who pointed out that the proportions of human limbs are much more similar to monkeys than apes. Humans have relatively short arms in comparison to their legs, like monkeys and unlike apes. Thus it is likely that the limbs had been used in a similarly monkey-like fashion in human evolution.

The tide of discovery over the years has tended to support the knuckle-walking side of this debate over the arboreal quadrupedal side. Molecular evidence that humans and chimps are very closely related, discovered by Vincent Sarich and Allan Wilson at Berkeley in the late 1960s, was taken as tacit support of the knuckle-walking hypothesis. Fossil discoveries of early hominids were described that clearly showed strong hand and foot grasping abilities, comparable to chimp climbing and brachiational anatomy, and distinctly different from monkey-like quadrupedal adaptations. But, conversely, there was little or no evidence from the fossil hand bones that early hominids had any history of knuckle-walking. And one of Washburn's graduate students, Russell Tuttle, in a detailed comparative study of ape and human hands, found no support for a knuckle-walking phase in human ancestry. So there the matter has rested: a confused idea of where hominid bipedalism might have arisen.

Grappling with two bones that I discovered at Sahabi, I became involved with the issue of the origin of hominid bipedalism. The magnitude of the questions that these bones might answer is only matched by their poor state of preservation. Thus all conclusions that I have drawn from them have been very constrained. Nevertheless, the specimens have allowed me to frame a new hypothesis on the origins of hominid bipedalism that can be tested with future discoveries at Sahabi and other sites in the 5-to-10-million-year-old time range.

The more important of the two bones from Sahabi is a fibula, the outer bone of the lower leg. The name comes from the Latin for "clasp or buckle," which the fibula forms with the tibia from the other side to form the ankle joint. The ankle joint is the keystone of an arch that transfers all the body's weight to the supporting foot in a upright-standing hominid. The surface of the joint is oriented parallel to the ground. In apes the ankle joint is oriented much differently. It slopes down and out so that the foot tends to angle in. This allows an ape to hold onto a vertical tree trunk with its foot grasping the trunk. The fibulas in hominids and apes, then, are very different. In hominids the fibula is straight and has a strong shaft because the bone is forming a part of the ankle joint and transferring weight straight up through the leg. The ape fibula, on the other hand, has a heavily built head because it is supporting the outside part of the ankle against gravity. The shaft of the ape fibula is quite thin because most of the body weight is transferred by the larger tibia on the inside of the leg.

The Sahabi bone looks like a human fibula. It is missing its upper half and the part that is preserved has been chewed by some carnivore. Yet it has a small head and robust shaft, like a hominid fibula. If it is a hominid fibula, its muscle attachments indicate that the flexors of the foot were quite strong. In fact, the bone is so dense that no marrow cavity can be seen halfway up the shaft where it is broken. If it is a hominid fibula, the indication is that the ankle joint was substantially similar to that of later hominids at the very early date of 5 million years ago for Sahabi. I published the specimen as a suggested hominoid fibula in 1987, but the bone is unusual and because of the incompleteness of the specimen I am not sure of its attribution.

The other bone from Sahabi that may be relevant to the bipedalism issue is the infamous clavicle, or collar bone. Clavicles are the most frequently misidentified bones in the hominoid fossil record. Supposed hominoid clavicles from the Early Miocene site of Rusinga in Kenya and supposed hominid clavicles from Olduvai Gorge in Tanzania were later shown to be reptile limb bones. Supposed early hominid clavicles from one of the South African cave sites turned out to be lateral toe bones of the three-toed horse, *Hipparion*. Sahabi added a wrinkle to this parade of bloopers and misidentifications in the fossil record—if it is a misidentification. Tim White concluded that the supposed hominoid clavicle from Sahabi was a posterior rib of a dolphin. I have published two papers in which I suggest that it is more likely a clavicle. But again the outside half of the bone is missing, again probably having been chewed off by a carnivore, and conclusions must remain tentative until more complete remains have been discovered.

What is more important here is how the Sahabi specimens changed my thinking about how early hominids may have begun to walk upright, rather than whether either of the bones supports these ideas. The specimens from Sahabi forced me to look at the assumptions in the hominid bipedalism debate. I found that many of them are unwarranted, based on a gossamer of evidence even more insubstantial than the little chewed-up bones from Sahabi.

The primary assumption that I found likely to be incorrect is that the mode of locomotion that we see in the modern great apes is primitive. In other words, if this assumption were correct, we would expect the common great ape-hominid ancestor to show the locomotor patterns of the great apes. This viewpoint, fundamentally, is expressed by the knuckle-walking hypothesis, which says that the common ape-hominid ancestors were knuckle-walkers because chimps and gorillas are knuckle-walkers today and this is the primitive pattern. This assumption is no longer supportable.

New molecular evidence based on DNA comparisons of the living apes and humans overwhelmingly supports the fact that gorillas split off from the hominid-chimp lineage first. This finding means that either the common ancestor of gorillas, chimps, and humans was a knuckle-walker or that the gorilla and chimp evolved knuckle-walking independently. The earliest hominid fossils as well as modern humans do not indicate any trace of knuckle-walking heritage, and the orangutan, an outside group, is not a knuckle-walker (it fist-walks). Thus there is no support for the claim that knuckle-walking was the primitive method of locomotion in the ancestor of the great apes and hominids. Knuckle-walking seems more likely to have arisen independently, in parallel, in the chimp and gorilla lineages.

What could have been the primitive locomotor pattern? Certainly not fist-walking, which apparently occurs only in the orangutan. Could it have been climbing, with bipedalism when on the ground, somewhat like the small-bodied gibbon? What other clues might throw light on this question?

## BODY SIZE AND BIPEDALISM

One clue about the locomotor behavior of the common ape-hominid ancestor is its body size. At a conference in Paris, the African mammalogist Jonathan Kingdon from Oxford University made the almost offhand observation that the first bipeds would have to have been small-bodied. He noted that bipedalism is a very rare form of locomotion among mammals, but when it occurs it tends to be accompanied by high metabolic activity and small body size. From a theoretical viewpoint, then, he would bet that the earliest bipeds were small-bodied.

This perspective fascinated me. In the history of paleoanthropology there has been a long debate over whether the missing link to the apes and the earliest hominid ancestors had been giants or pygmies. The eminent German-American paleoanthropologist Franz Weidenreich had advanced the *Giganthropus* theory of human origin. Weidenreich had described the famous *Homo erectus* Peking Man fossils and, puzzling over their many unique cranial anatomical characteristics, ascribed the heavy brow ridges, the unusually thick cranial bone, the massively built back parts of the skull, and the unusual keel running along the top of the skull to remnants of descent from a much larger, giant form of human. Some fossils from Java named *Meganthropus* and some large molar teeth from China originally termed *Giganthropus* and now known as *Gigantopithecus* were cited by Weidenreich as possible support for his theory. But the discovery of the australopithecines with cranial anatomy very much different from Weidenreich's prediction pretty much finished off the *Giganthropus* theory of human origins.

On the other extreme was the pygmy theory of human origins, proposed in 1905 by the German anatomist Julius Kollman. One of Kollman's contemporaries disparagingly referred to it as the "dwarf" hypothesis (Zwerg in German). Kollman had noticed that the skull form of adult humans is quite similar to that of juvenile apes. Making the theoretical jump from growth to evolution—from ontogeny to phylogeny—Kollman hypothesized that the missing link had been a small-bodied form with much more rounded skull features than either the modern great apes or fossil *Homo erectus.* He thus drew the wrath of not only the paleoanthropologists who (rightly) put *Homo erectus* on the human family tree, but also the comparative anatomists who were very confident (rightly) in their belief of a close connection between the modern great apes and humans. Because the paleoanthropologists and comparative anatomists with opposing viewpoints carried the day, Kollman's theory is largely forgotten today. Jonathan Kingdon's comment at the Paris meeting recalled to my mind Kollman's ideas and I began to think that perhaps Kollman was not so wrong after all.

Interestingly, the fossil record of the earliest hominids and related hominoids and the perspective of comparative primate anatomy eighty-five years later lend support to Kollman's idea that the common ape-hominid ancestor was small-bodied. The fossil hominid footprints at Laetoli, dated at 3.6 to 3.8 million years ago, correlate in size to the feet of people only three-and-a-half to four feet tall. The earliest and most complete skeleton of an early hominid is Lucy from Hadar, about 3 million years old, and she is reconstructed to be about four feet tall. Other individuals at Hadar would have been larger, however, judging by their teeth and other body parts. Nevertheless, it is fair to say, and now quite uncontroversial to conclude, that hominids evolved from small-bodied ancestors.

What about the ancestors of the great apes? Did they come from small-bodied forms as well? Anthropologists had assumed for quite some time that the large body size of the living chimp, gorilla, and orangutan was the primitive, ancestral condition for the common ancestor. But now we have reliable molecular evidence that the chimp and gorilla lineages evolved a number of characteristics in parallel. Knuckle-walking, thin molar enamel, and large body size were likely three of the major characteristics that evolved independently in the chimp and gorilla lineages. Observations made by Adrienne Zihlman at the University of California Santa Cruz that the smallest of the chimps, the pygmy chimp or bonobo, was closest anatomically to the common ape-human ancestor supports this idea.

There are now some skeletal remains from Pakistan that relate to the ancestry of the orangutan. These are fossil bones of the ape *Sivapithecus*. A facial skeleton discovered by David Pilbeam of Harvard University in the Late Miocene rocks of the Siwalik Hills of Pakistan at about 8 million years ago showed for the first time that Sivapithecus had an indisputable anatomical similarity to the modern orangutan and was likely ancestral to it. Notably, first, a relatively intact upper arm bone, the humerus, does not show the specialized anatomy associated with fist- or knuckle-walking seen in modern orangutans. Second, the fossils of the limb bones of *Sivapithecus,* although not associated with the skull and tooth fossils, reveal a range of size much smaller than the modern orangutan. Extrapolating to whole body size from fragmentary limb bones, the estimate is that Sivapithecus was about the size of a medium-sized monkey or a gibbon. Thus the evidence from the Asian branch of the great apes does not contradict the idea that the body size of the missing link was small.

The fourth branch of the apes, after chimps, gorillas, and orangutans, is that of the lesser ape, the gibbon. The gibbon is the smallest of the living hominoids, and in interesting support of Kingdon's suggestion, it is bipedal when on the ground. The gibbon is unique in having evolved very long arms, and it is a superb high forest acrobat, swinging from tree branch to tree branch with amazing speed and agility—unlike our conception of the locomotion of the missing link. But in other ways it is likely the closest living analogue to the common ancestor of the modern apes and hominids.

If the missing link were small-bodied, then could we conclude that the ancestors of the great apes might have been bipeds? This, I thought, was a radical new idea: that knuckle-walking apes had descended from hominid-like bipeds! Some library research showed, however, that this exact idea had been proposed in 1897 by the British anthropologist Robert Munro. He had suggested that erect posture was an ancient adaptation and that apes had only lately "degenerated" into a quadrupedal form of locomotion. But how would the idea work with new data? Could I come up with a scenario of hominid origins that made sense?

## SCENARIOS OF HOMINID ORIGINS

Building scenarios and models in hominid evolution is danger-ous business, mainly because it is so difficult to bring in all the elements of the action at the right times and in the right amounts to accurately represent what really happened in nature. And such hypotheses are difficult to test. The anthro-pological literature is replete with "just-so" speculations, so named because of Rudyard Kipling's *Just So* stories like "How the Rhinoceros Got His Skin" and "How the Elephant Got His Trunk." When an anthropologist finds new evidence, he or she may postulate, not surprisingly, that a certain model of evolu-tion made it just so. The difference between these speculations and hypotheses is that hypotheses have to be empirically verifi-able, testable, and disprovable.

I began the building of a scenario for the evolution of bipedalism and the initiation of the ape-hominid split by recon-structing the environment and the ecology of its time and place. Evolution is a complicated dance between environment and life: changes in one affect and effect changes in the other. But the environment was here first, and its large-scale changes, like weather and climate, occur normally with little or no input from living systems. The environment, then, is the starting place.

If the anatomy of the chimp and the gorilla had changed sub-stantially from the ape-hominid ancestor, one thing that almost certainly had remained the same was their habitat. We could not necessarily rule out the possibility that chimps and gorillas had left the forests, evolved in the woodlands and savannas, and then returned to the forests, but it was a more complicated scenario than necessary. Most likely the environment of the apes had remained the same since the Early Miocene, 18 million years ago—forest. This, then, is the environment of the ape-hominid common ancestor: the largely unknown forest habitat of the missing link.

It goes without saying that the forest habitat of the common ancestor had to be in Africa. This is where the chimp and gorilla live, and there is no evidence that they ever extended their range north of the Sahara into Eurasia. The time of the common ancestor has to be around 6 to 12 million years ago. This is the time frame indicated by the late divergence hypothesis for the gorilla, chimp, and hominid splits from common ancestral populations.

But where in Africa? East Africa by 13 million years ago was already an open savanna woodland. One or two isolated teeth and jaws have been found at sites dating 7 to 10 million years ago in East Africa, but I did not think that these fossils had much, if anything, to do with the ape-hominid split. If I was correct, our ancestors and those of the chimp and gorilla were still living in forests at that time. The apes named *Kenyapithecus*, *Afropithecus*, and *Turkanapithecus*, discovered and named by the Leakeys in East Africa, all would have been parallel side branches of an earlier radiation of apes. Monte McCrossin and Brenda Benefit of Southern Illinois University have reported the recent discovery of the arm bone of *Kenyapithecus* at Maboko Island, Kenya, which shows anatomy much different from that of living apes. The bone suggests that *Kenyapithecus* was not a climbing or hanging animal and was not closely related to hominids or modern apes.

*Kenyapithecus* and related apes in middle to late Miocene East Africa could have been left over from the breakup of the East African Early Miocene forests. When East Africa became even drier as the Miocene and Pliocene advanced, the savanna woodlands shrank even more and these relic species probably became extinct without issue.

The most conservative and species-rich forest area of Africa looks like unbroken green carpet from the air. It is known as the Central Forest Refuge because during periods of climatic aridity during the ice ages this area of eastern Zaire remained forested even though surrounding forests gave way to savanna and even desert. By contrast, during the last maximum ice age cooling, about 18,000 years ago, sand dunes from the Kalahari Desert in southwestern Africa extended up to the southern margin of the Zaire River in what is now full-blown rain forest. This deforestation did not happen in eastern Zaire, where high mountains caused rain to drop along the western margin of the western Rift Valley and thereby maintained the dense forests despite dry conditions elsewhere. The Central Forest Refuge has harbored such ancient forest species as the okapi, a short-necked giraffe that looks very much like the 40-million-year-old fossil ancestors of the familiar long-necked variety; the pygmy buffalo, which lives in cloud forest in the Ruwenzori Mountains; and the needle-clawed galago, a large-eyed nocturnal prosimian primate found here and nowhere else.

There are two other large forest refuges in Africa, characterized by unique forest species endemic only to those areas. They are in the far west of Africa, and they harbor such species as the mandrill, a baboon with a multicolored variegated facial ornamentation, and the pygmy hippopotamus, smaller and less committed to a life in the water than the familiar *Hippopotamus amphibius*. But the Central Forest Refuge has the largest number of endemic species and is the largest in area. Most important to anthropological concerns, the central refuge contains two of the three subspecies of the gorilla (the mountain and the eastern lowland) and the chimpanzee. One of the western refuges lacks gorillas entirely. The enigmatic pygmy chimp or bonobo, which lives in the forested region of Zaire that was quite arid during the last glacial period, also must have come from the Central Forest Refuge; the other forest refuges were too far away to account for the bonobos that recolonized the forests south of the Zaire River. For these reasons, I thought the evidence of modern species distributions pointed toward the Central Forest Refuge as the source for the ancestors of the African great apes and hominids—the environment of the missing link.

Why would the ancestors of hominids have left the forest?

What would have caused them to venture out on the hot savanna with few shade trees and few trees to climb for protection? I thought the answer was ecological. In a scenario that must have been played out thousands upon thousands of times over many generations, a group of prehominids, faced with dwindling fruit and food sources in a shrinking patch of forest, made the trek to another distant patch of forest. The trek was hard, and the prehominids ate what they could in the savanna until they arrived back in a familiar forest habitat. A number of the group could not keep up; they became weak and were picked off by carnivores or succumbed to hunger. The prehominids who could eat any of the tough-skinned fruits, nuts, roots, plant leaves, or small animals of the savanna were favored by natural selection. When the prehominid group arrived again in optimal ecological conditions, the individuals who had survived the savanna trek were the ones left to reproduce.

This first step away from the savanna had resulted, I thought, in the initial split of the great ape-hominid group: the ancestors of the gorillas had stayed at home. Through ecological chance, those hominoids whose populations had remained surrounded by forest never adapted to the edge environments that were likely to form isolated forest patches surrounded by savanna. Mountains were the most consistent places in Africa for forests to retreat in times of aridity. The two forest refuges in Africa with large mountain massifs, Cameroon and eastern Zaire, with its Virunga volcanoes and Ruwenzori regional uplift, are where the gorillas are found today.

The chimp-hominid ancestor had remained a lowland forest species, subject to periodic bouts of savanna life. Chimps today can live in quite marginal forest habitats. In Sierra Leone, for example, they range into savanna areas for food resources. The diet of the chimp encompasses such savanna delicacies as termites, small antelopes, and nuts, which some chimp groups have been observed to crack open with rocks. The gorilla, on the other hand, cannot eat anything outside of a restricted regime of succulent forest vegetation, composed of tender shoots and the soft pith of trees. Evolving large body size as a response to increased predator pressure in forests of reduced size and possibly also as a result of the frequently cold, wet conditions in montane habitats, the gorilla fell forward on its knuckles as its upper body weight became too great to sustain in an upright stance. This knuckle-walking adaptation then evolved as a particularly effective terrestrial means of locomotion. In Zaire I realized the advantage of four-appendage, low-center-of-gravity locomotion when I visited the slippery, vine-covered, hanging-moss habitats of the mountain and eastern lowland gorillas. As a biped, the experience of lurching, slipping, falling, and crashing through the undergrowth to emerge upon a serene group of feeding gorillas who move quietly, effortlessly, and rapidly through this terrain makes this point more forcibly than reading any number of scientific treatises on the biomechanics of knuckle-walking and bipedalism.

## CLIMATIC CHANGE AND APE-HOMINID EVOLUTIONARY SPLITS

It should be obvious by now that the missing link has yet to be clearly pinned down. We believe that gorillas split off first, chimps second. I have argued that these splits could have occurred in a number of places in Africa—not just the east where anthropologists from Louis Leakey to Don Johanson have done so much work. Now I have suggested a *reason* for the gorilla split: climatic change from forest to savanna.

No fossil sites record this first split away from the common African great ape-hominid ancestor. Its timing, however, can be surmised by recourse to data from an unlikely place—the deep sea. The Deep Sea Drilling Project was an international research undertaking that yielded thousands of feet of cored sea-floor sediment from dozens of ocean voyages around the world. I have been to the repository of these cores at Lamont-Doherty Geological Observatory, north of New York City, and they look just like long tubes of mud of differing colors and textures. But the perspective they provide on the history of the earth's climate is unprecedented. By analyzing the varying ratios of heavy to light oxygen molecules in the sediments and in the shells of small animals found in the cores, scientists have been able to tell what the temperatures of the oceans were. During colder times, heavier oxygen molecules with two extra neutrons, the $O^{18}$ isotopes of oxygen, become enriched in the world's oceans because the lighter isotopes of oxygen have become trapped in greater proportion in land ice sheets. Ocean temperatures, of course, are closely related to air temperatures, and cold temperatures in the tropics generally mean aridity. Aridity means shrinking forests.

Beginning after 11 to 13 million years ago, in the middle Miocene, the deep sea core records a significant decrease in world ocean temperature (see chart). The cause of this decrease in world temperature is the increase in size of the Antarctic ice sheet. Cold ocean currents began flowing up from Antarctica along the west coast of Africa, and they decreased the amount of moisture that the air could hold. Rainfall decreased over tropical Africa. This decrease in rainfall can very likely be implicated with the environmental changes accompanying the gorilla–chimp-human split.

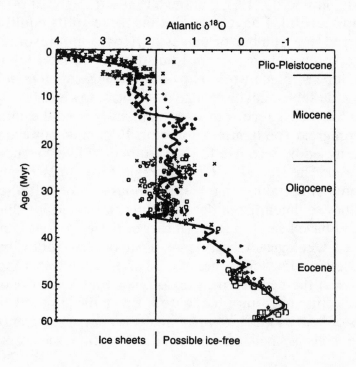

Decreased rainfall caused forests to shrink and increased the distances of open savanna between patches of trees over large parts of Africa. Forests receded up mountains. The ancestors of the modern gorillas were split off from the ancestors of the chimps and the hominids.

Following this scenario, what would have happened during the chimp-hominid split a couple of million years later, if we accept the late divergence hypothesis dates? In this the second of the great ape-hominid splits, it was the chimps who stayed at home in the lowland forest and forest fringe environments and the hominids who set out on a new environmental course. Chimps evolved large body size, possibly as they were subjected to greater predation in the shrunken forests, and in parallel to the gorillas they fell forward onto their hands to support the increased weight of their upper bodies. The chimps and gorillas still retain the primitive, proportionately greater development of the upper limbs of climbers and hangers, and with increasing body size this proportionate difference is exaggerated. In contrast, as hominids later evolved larger body size, they reduced the relative size of the upper limbs and upper body. Their legs evolved relatively greater length and were used increasingly in terrestrial locomotion. The chimps were unable to pursue a life entirely on the savanna because they were not able to traverse the long distances involved to find food and sheltering trees.

Natural selection solved the problem of shrinking forests in a different way for the hominids. Hominids stayed small and retained their ancestral bipedal way of getting around. They escaped the forests with their now highly concentrated populations of predators and survived by running away—not fast, just far. The biting jaws and slashing claws of natural selection were always nearby to weed out the tired, weak, and infirm, but especially the imperfect bipeds. The savannas held vast food resources, but these resources were more sparsely scattered than in the forest. Hominids evolved a high-efficiency, long-distance locomotor system, bipedalism, to meet the challenge that the much expanded sizes of their home territories provided.

What forces of nature effected the evolutionary divergence of the chimps and the hominids? Something must have happened that finally severed the connection to the forests for the population of prehominids that became cut off from their prechimp relatives. I think that there are two possibilities, and I set out to test both of them.

---

The first scenario involves Sahabi and relates to the drying up of the Mediterranean Sea during that geological period between about 5.0 and 6.3 million years ago known as the Messinian Event. When virtually all of the water was removed from this vast expanse of territory, there must have been dramatic effects on the precipitation over the whole of North Africa as well as the entire area of Eurasia surrounding the Mediterranean. Rainfall must have decreased precipitously, because there was little or no evaporation from the hypersaline pools of water left in a few depressions in the Mediterranean basin. A number of authors have speculated that the climatic effects of the Messinian Event had much to do with the emergence of hominids, but no one has tested the idea.

Sahabi is the perfect field laboratory to test the hypothesis that the Messinian Event had a major climatic effect on hominid emergence. It sits atop a ten-meter thick stratum of gypsum that was deposited at the bottom of a remnant of the Mediterranean, much saltier than the modern Dead Sea. Because of the immense thickness of this bed, it has to be the geological signature of the Messinian Event, also found undersea by the Deep Sea Drilling Project. The fossils that we discovered at Sahabi then were covered by waters that refilled the Mediterranean basin about 5 million years ago. They record what the fauna and flora were like after having undergone over a million years of arid conditions, as indicated by deep sea core record.

This decrease in rainfall, a result of decreasing global temperatures during the middle to late Miocene Epoch, accounted for the spread of open habitats—savannas and savanna woodlands—over much of East Africa. This period witnessed the first big break-up of the Early Miocene tropical African forests. Faunas adapted to savanna habitats evolved in eastern Africa and included apes that the Leakeys have found, particularly

*Kenyapithecus*. In my new scenario I propose that *Kenyapithecus* is too early to have anything to do with later hominids and African apes. *Kenyapithecus* got caught in one of the dwindling forests of the late Miocene and became extinct.

The deep sea core record of global temperature change shows a continuing trend toward a cooler climate and hence generally drier conditions in tropical Africa. As part of the same environmental shift toward progressively greater aridity that snuffed out *Kenyapithecus* in eastern Africa, the ancestral chimp-hominids in central Africa were isolated from their proto-gorilla cousins in the forest refuge. This was the split between the gorillas and the chimp-hominids. It occurred about 11 to 13 million years ago according to the late divergence hypothesis and the record of global climate change.

How might emergent hominids fit into this picture? One scenario would be that a group of Central African populations living at the forest edge became stranded north of the Central Forest Refuge by encroaching arid conditions. They were sustained by local water sources such as rivers flowing from the south or by lakes fed by these waters. Undergoing the same selection of shifting savanna and forest mosaics that their ancestors had undergone earlier to the south, they became adapted to the vast savannas that occupied what is now the Sahara Desert. The paltry limb bone fossils that I had found at Sahabi might well indicate the presence of this hypothesized hominid ancestor.

The test of this scenario is at the fossil site of Sahabi. By using every possible research tool, we sample and analyze the past environment, and by using every possible controlled collection technique we gain knowledge of all aspects of the past flora and fauna. Only then do we put the hominoids that happen to be found during the faunal collection into the scenario. It is very important to avoid a treasure-hunters' approach, the Indiana Jones Technique, of searching only for one object, be it a crystal idol or a 5-million-year-old hominid skull. Should that object be found without its envelope of surrounding information relating to time and environment, its context, its usefulness in testing scientific hypotheses is minimal.

Near the end of the last field season at Sahabi I collected a skull bone of a primate that had the globular shape of an ape. It was small for an ape, about the size of a gibbon, but it was lower and flatter. An aberrant monkey? An early hominid? Did it go with the same species of animal to which the fibula and clavicle belonged, if they belonged together? No one knows, nor will they ever, unless we can get back to Sahabi to reinvestigate the fossil beds, many of which have never been surveyed. Sahabi is also the best site known at present to test a North African scenario for the chimp-hominid split. Dating to just after the Messinian Event, it is situated ideally in time and place to test whether hominids first diverged to the north of their ancestral forest homelands.

The other scenario for chimp-hominid divergence focuses around another cataclysm of earth history, the formation of the African Rift Valley. There are actually two Rift valleys, and they are so huge that they can be seen from the space shuttle. The Eastern Rift extends up from southern Tanzania to Kenya through Ethiopia and into the Red Sea. The Western Rift diverges from the eastern in Malawi and goes up between the borders of Zaire and Uganda. The Eastern Rift is the more impressive in terms of high faulted scarps, but the Western Rift is easier to follow on a map because it provides the basins for the African Great Lakes—Tanganyika, Kivu, Rutanzige (Edward), and Mobutu (Albert)—as well as determining the initial course of the White Nile. The Eastern Rift has received most of the press over the years, first by geologists attempting to figure out how such large-scale, steep-sided grabens formed, and then by paleoanthropologists who began making significant discoveries of fossil hominids in the sediments accumulating in the subsiding basins. The geologists eventually figured out that the East African Rift Valley system was a rare example not located on the sea floor of the spreading of plates of the earth's crust that accounted for the drifting positions of continents. The African rift valleys had stopped drifting apart, at least temporarily, and for that reason were not under the sea. Recent redating of the Eastern and Western Rifts indicate that the eastern started about 14 million years ago, and the western only about 8 million years ago.

Paleoanthropologists have always thought that the Rift Valley had something to do with hominid origins. They have almost always concentrated on the Eastern Rift, probably because it is more accessible, not covered in vegetation, and extends though countries where English is spoken, more important to British and American fieldworkers than many would think. But I was fascinated with the Western Rift, more remote, more unknown, and probably much more important to scenarios of earliest hominid origins than the Eastern Rift.

The Western Rift is one of the great dividing lines of Africa. Its upfaulted and uplifted shoulders have caused mountains to form. The Ruwenzori Mountains are the tallest in Africa, after the Eastern Rift mountains Kilimanjaro and Kenya, but virtually the entire western rim of the Western Rift is a region of uplift. This uplift traps rain clouds on the western flank of the Western Rift and makes the rift valley floor much more arid that the forests to the west. The rift becomes a major delimiter of species of forest plants and animals.

Rivers have played a major part in the natural history of Africa, and the Western Rift affects the continent's two largest, the Nile and the Zaire (or Congo). Uplift of the Western Rift turned waters that were flowing into the Nile to the west and formed the headwaters of the Zaire. Downfaulting of the Western Rift basin formed the conduit along which the headwaters of the Nile flowed. Rivers not only serve as routes for the aquatic animals that live in them but also as delimiters of species that cannot cross them and as foci for gallery forests that surround them. Changing water courses through time have had profound effects on the geographic distributions of species.

My second scenario of hominid-chimp evolutionary divergence posits a major role for the Western Rift. As the Western Rift began to form about 8 million years ago, previously continuous forests extending from the Central Forest Refuge into eastern Africa were broken up. The edge-living chimp-hominids were subjected to more tenuous forest connections back to the central refuge. Eventually some of the eastern populations were cut off, becoming isolated genetically from the populations that stayed on the western side of the rift. The eastern populations stayed small and became the long-distance, bipedal hominids, inhabiting the expanding savannas of eastern Africa, and the western populations became the chimpanzees, inhabiting the forests and the forest fringes of Central and West Africa. Only when the steep escarpments of the Western Rift were lowered by millions of years of erosion and when increased rainfall allowed forest extensions into the east, as in the present, have a few chimp populations extended east of the Western Rift. Jane Goodall's chimps at Gombe Reserve on the eastern border of Lake Tanganyika are one such example.

This Central African scenario has the advantage of a clear biogeographic pattern of support. Many fossil hominid sites are known in eastern Africa, but none are known west of the Western Rift, partly, of course, because of the dense forest and the lack of any significant exploration. But in the known fossil sites in eastern Africa and in related deposits in southern Africa where early hominids are found there are no apes. This absence implies that the east belonged to the hominids and the west belonged to the apes. Modern distributions of the gorilla and the chimp stop for all practical purposes at the Western Rift, and this is likely to represent an ancient line of demarcation.

The main weakness in the Rift Valley scenario for hominid divergence is a lack of explanation for the increase in size of the chimp. I mean by this more than just the growth in body size allowed by an optimum diet. There was an evolutionary change in the chimp's adaptation to larger body size. If there was not progressive desiccation of environments, as in the North African model, but a simple cutting of the forests by the rift, there would have been be no greater predator pressure or food-related pressures to drive the evolution of larger body size. Were other selective pressures involved that I have not thought of? Perhaps competition from monkeys became severe, or the aridifying effects of the Messinian Event had an impact even in Central Africa. There was no way to answer the questions without more data.

My test of the second scenario of hominid-chimp divergence has been and continues to be the Semliki Research Expedition. In the 1950s fossil deposits were found by Belgian geologist Jean de Heinzelin on the slopes of the western wall of the Western Rift in eastern Zaire. The deposits were thought to extend in age from perhaps 20 million years ago to the Pleistocene Epoch, which ended only 10,000 years ago. There were fossils, but how productive they were was hard to tell. De Heinzelin and his crew had to leave Zaire abruptly in 1960 without even making a full reconnaissance of the deposits.

When my ongoing test of divergence hypothesis 1 was abruptly halted by the severance of U.S. diplomatic ties with Libya, I decided to test divergence hypothesis 2 in the Western Rift of Zaire. U.S. State Department authorities upholding executive orders of President Ronald Reagan would not let me fly into Libya to get my expedition Landrover to transport it to Zaire. They would not even let me go the border between Libya and Tunisia to get it if my colleagues at the university would drive it there for me. A year later, in an apparent attempt to assassinate Libyan leader Qaddafi, American jets dropped bombs less than a mile away from Garyounis University where our lab was with most of the Sahabi fossils. I talked with my good friend and colleague Wahid Gaziry by phone in Benghazi, and he was very saddened by the attack. Luckily, no one at the university had been injured, and the buildings had not been hit. But Wahid managed to get the papers together and get the Landrover on a boat bound for Antwerp, where I arranged for it to be shipped to Matadi, Zaire. I started preparations for a new chapter of research.

# 7

# The Other Rift Valley

### THE CRUCIBLE OF HUMAN ORIGINS

The scientific establishment, like any establishment, can be remarkably intransigent when it comes to new ways of looking at things. In my case it was looking *for* things, fossils of ancient hominoids to be exact. One of the peer reviewers who had some claim to knowing about field conditions in Africa flatly stated that my proposed fieldwork in the Western Rift Valley of Zaire was "unfeasible." The grant proposal was turned down. But I was finally able to convince the Wenner-Gren Foundation in New York that the project had some merit and would not be a waste of money. With this grant and funds from a New York University Research Challenge faculty grant, I had enough to make a foray into the fossil sites of eastern Zaire for the first time since they had been discovered over twenty years before.

Bunia, Zaire, June 1982. The Air Zaire plane set down on the runway that appeared from the air to be in the middle of nowhere. It had not been an easy trip from Kinshasa, Zaire's capital, that now lay 1,000 miles to the west. We had started for the airport from our hotel at 4:30 that morning. The plane, which was scheduled to take off at 7:00, did not even board until 10:00, when there had been a mad rush to the plane. There was no explanation for the delay. This was just the way things were done here. At noon we landed at Kisangani, formerly Stanleyville, a name that evoked images of the massacre and civil war that had rocked Zaire in the 1960s. There we were to transfer to a Fokker prop plane, but as we were beginning to board, the twenty passengers were turned back to the terminal. After a few minutes, one of the Air Zaire officials came up to me, addressed me as "patron," and announced in very proper French that there was a mechanical difficulty with the plane. I thanked him and sat down to wait, all the while pondering why I of all the passengers should be singled out to be told that there was a mechanical problem. After half an hour when there had been no sign of any mechanics working on the plane, I went and asked if the official thought we might make it to Bunia that day. He didn't know. Then something clicked. I pulled out a 20 Zaire note, about $1.50 at the time, and gave it to the official, suggesting that this might help the mechanics in their work. He thanked me, immediately departed for a back office, and in fifteen minutes we were airborne on our way to Bunia.

When we arrived, Bunia did not hold out a hospitable welcome. A drenching rain began as we emerged from the airport, there were no taxis in town, and neither of the two hotels had room for us. But we were here, near the fossil sites, and nothing was going to keep us from getting into the rift to see them.

## THE LOWER SEMLIKI

Jean de Heinzelin had told me about his discovery in 1960 of very promising and very old fossil sites in the most inaccessible reaches of the Western Rift. The deposits were located on the rift shoulder in the northern part of the Semliki River valley (termed the Lower Semliki because the river flows north) and along the southwestern shore of Lake Mobutu, formerly named Lake Albert. De Heinzelin had fled the area for his life

in 1960 during the revolution that brought Zaire its indepen-
dence from Belgium. Having returned to the valley from his
base to the south to pay his workers, despite warnings of
impending unrest, he had been captured by a group armed with
spears and knives that included some of his crew. Under the
influence of *pombe*, a local banana beer, and *bangi*, locally
grown marijuana, the men had forced de Heinzelin at spear
point to drive them around the area and demanded all his
money. Feigning car trouble by flooding the engine, de
Heinzelin had made his escape when all the men had left the
midday heat of the car to sleep under a tree. His last view of
the northern Semliki valley had been out of his rear-view mir-
ror as an angry group of men ran after him brandishing spears.

De Heinzelin referred to the area by the names of two small
rivers that ran off the rift wall—Sinda-Mohari. When whispered in the circles of the cognoscenti of African paleoanthropology, the name had a far-off, exotic sound that elicited the
same emotions as the names of El Dorado or Shangri-la had to
earlier generations of explorers. There was the same sense of
great wealth waiting to be discovered, in this case hidden
knowledge in the form of fossils that the earth had successfully
kept secret for millions of years. And the same sense of danger.
The wealth might be there, but would the cost of obtaining it
be too great? Most modern researchers did not want to risk
their lives.

I thought the fears had been greatly exaggerated.
Missionaries, teachers, and doctors had been peacefully working in eastern Zaire for two decades. They had been able to
find fuel and spare parts for their vehicles and maintain good
relationships with the people of the area. I did not see why we
could not do the same, although I was under no misconcep-
tions about how difficult this undertaking was going to be.
There was still unrest in the southeastern part of the country,
in copper-rich Shaba Province. It was from here that Zairean
rebels had come from across Lake Tanganyika to attack Jane
Goodall's chimpanzee research station at Gombe, Tanzania, in
the early 1970s. Three graduate student researchers had been
abducted to Zaire and were released only after the payment of

a reputed $1 million ransom raised from the United States. And Dian Fossey had first attempted her gorilla studies in eastern Zaire, but had been abducted from her camp by Zairean soldiers, taken to the Rwandan border, and released.

My strategy was simple. The Lower Semliki area was peaceful and was controlled by the authorities in Kinshasa, the Zairean capital, as well as by the local chief of the Walenda-Bindi tribe, now termed a *collectivité* in modern Zaire. I would request letters of authorization from both the authorities in Kinshasa and from the Walenda-Bindi chief to allow us to go into the areas. De Heinzelin said that he could give us names of Catholic missionaries in the area who could help us with logistics. If the authorities would not give us authorization or could not guarantee our safe passage, then we would have to turn back. If we could get no logistic support from the missionaries then we would have a hard time covering very much area in surveying.

When we arrived in Bunia I came armed with official letters emblazoned with the symbol of Zaire, an upraised hand holding a lighted torch. The letters named the members of our group and authorized us to undertake the field survey. There were three other members of the group: my former wife, Dody, and NYU graduate students Bill Sanders and Monte McCrossin, both in Africa for the first time. I also had names of missionaries in Bunia that de Heinzelin had sent me from Belgium. When the two hotels in town turned us away, we headed for the mission in the hope that they might help us or at least give us a room for the night. When the missionaries heard where we intended to go, any assistance in transportation was out of the question. They just looked incredulous that I would even think that they would lend something so valuable in this country as a working vehicle to a total stranger. In retrospect I suspect that the missionaries thought that we were gold prospectors. Gold had been discovered in several of the streams and rivers around Bunia, and fortunes were being made by traders in gold, mostly Greeks and Malians.

When we were turned away from the mission, we headed for the only other haven available, a small African hotel with mud walls and dirt floors in the *cité*, or old section of Bunia. Its name was the Africa Pole-Pole (*pole-pole*, pronounced poe-lay poe-lay, in Swahili means slowly). But things were anything but slow this Sunday night, and the excitement had just picked up with the arrival of four *wazungu* laden with packs, cameras, and equipment. I had been too exhausted to bargain, and I paid the hotel the exorbitant fee of $20 per room in the hope that the owner and his retinue would feel sufficient gratitude for my generosity to prevent our being disturbed. It was not a restful night. Trying to sleep on rope-mattress beds, we could hear a raucous crowd surge toward our two adjacent rooms only to be calmed by one and sometimes two voices. Only I could catch snippets of the Swahili and French debates about how much money the foreigners had and why they were here. Working in our favor were the official letter from the government in Kinshasa, the fact that I had paid the hotel manager so well, and that we had come from the mission, indicating that we were probably not mercenaries, probably in that order. In the end, nothing happened, but nobody slept that night.

In the morning the hotel manager informed me grandiosely that he had saved our lives numerous times during the night. I thanked him profusely, but politely refused to give him another $10 for this service, which I knew to be more than slightly exaggerated. It is very important to bargain in Africa. It is a primary way for people to make social contact and exchange opinions, and it even serves as a form of recreation. The manager asked me archly if we valued our lives so little. I said that we valued our lives very much but that I considered having paid him ten times the normal price of a room to be adequate compensation for his troubles. He laughed and we went into town to find breakfast, with our luggage pushed along on a hired *pouse-pouse*, a long-handled cart used in Zaire to transport everything from chickens to drums of gasoline.

We managed to set up a base of operations at the Hotel Semliki in Bunia, an establishment run by a Greek Cypriot named Andreou. Here we assembled food and supplies bought in town for the trip into the rift. I spent time trying to find a vehicle to rent or borrow, but with no luck. After days of waiting and negotiating, I finally got us a ride on a dump truck hauling concrete to the village of Gety, near our departure point by foot into the rift. The concrete was to be used to construct a dip trough for cattle tick deinfestation.

Our elation at finally leaving the Hotel Semliki for the rift was unbounded as we climbed onto the truck with another twenty or so riders. Andreou waved goodbye to us and assured us there would be room for us when we came back. His eyes said if we came back. The trip was slow and bumpy. We stopped at each stream so that the driver's assistant could jump off the back of the truck, open the hood, and fill up the radiator. He also filled a little long-necked tea kettle, for what purpose I could not imagine. By the time the truck had labored to the top of the hill past the stream its radiator was steaming and I realized why it had been necessary to fill the radiator. We then started down the next slope, crawling in first gear. The tea kettle came into play on curves when the driver had to apply the brakes. His assistant ran alongside and poured a stream of well-aimed water to cool off the smoking, worn brake shoes. At one point the driver stopped to fine-tune the diesel engine, which sounded fine to me. He called for a shovel. By placing the blade of the shovel in little well-worn notches on the engine nuts he loosened them and proceeded to bleed the diesel fuel to remove any air in the lines. He tightened the nuts by reversing the process and finished with a firm blow of his hand to the shovel handle.

The scenery was beautiful. Moving along at five to ten kilometers an hour at a good height above the roadway gave us a rare chance to see a part of Africa seldom visited by Westerners. It was green. Perhaps this was the major difference that I noticed between Zaire and the sere plains of East Africa. Much of the forest had been cleared by the Belgians years ago for pastureland, and there were still a number of cattle grazing along hilly slopes. Some long-forgotten Belgian travel agent had coined the terms "the Switzerland of Africa" and the "Route of Beauty" to refer to the region of Zaire and the road where we now found ourselves. It was cool because although we were at the equator the altitude on the rim of the Western

Rift was 2,000 meters above sea level. It is a little-known fact that the African rift valley has highly elevated sides as well as a down-dropped basin. We passed men in long overcoats and women wrapped in shawls against the morning cold and damp.

Each time we stopped for water and the deafening roar of the diesel was no longer in our ears, I noticed the far-off beating of drums. A lot of marriage celebrations I thought.

We passed two ancient signs, put up when this road was traveled by tourists. One noted that we were going from the drainage of the Congo River (now the Zaire) into the drainage of the Nile. The uplifted western margin of the rift wall separated the watersheds of the two largest rivers in Africa, one heading west, eventually into the Atlantic, and the other east and north into the Mediterranean. What a paradox I thought. This area that was now so remote from the modern world was the central biotic stage, the point of intersection of the forces that had molded hominid evolution. It was where the ancient forests met the savannas and where even the continent's great water courses diverged.

The next sign noted that Henry Morton Stanley had camped here for a year in 1874–1875 with Emin Pasha, the former German naturalist turned Islamic administrator of the remote province of Equatoria in the former British colony of Anglo-Egyptian Sudan. Emin had been isolated, at least in British eyes, when the Mahdi's forces had taken Khartoum and killed General Charles "Chinese" Gordon. Emin, however, had been much more ambivalent about being rescued. In any event, Stanley had made good use of the opportunity to extend his explorations and had discovered the Semliki River, into whose valley we were now heading.

We lumbered into a small town that seemed to serve as a stop for the trucks laden with dried fish traveling between Lake Mobutu and Bunia. Everyone piled out of the truck and headed for the shops and street vendors to buy something to drink and some lunch: roasted peanuts sold in increments of little metal cupfuls, loaves of French bread, and hard-boiled eggs. We were soon on our way again, going deeper and deeper into the bush. The road turned from a well-traveled graded dirt roadway to first a road wide enough for a single vehicle and finally to just two parallel wheel tracks. Our speed decreased even more as the driver meticulously picked his way through narrow and rocky stretches. Without a practiced eye, one might think the tracks were just dry stream beds, not a road that a large dump truck could or should pass through.

We arrived at the mission after dark, around 9:00 P.M. Our reception here was very different from the mission in Bunia. The missionary, a bright-eyed man in his mid-fifties with a Flemish accent, invited us into the mission for dinner. He knew that we were four and that one of us was a woman. I asked incredulously how he had known. "There are no secrets in Zaire," he replied. "The drums," I thought. Everyone was too tired for explanations, so the missionary showed us to immaculate rooms with a shower and bathroom down the hall.

In the morning over breakfast of tea, eggs, and toast, complete with local butter and jam made from the *maracuga* fruit, we told Pere de la Faille, the missionary, what we were doing. He was very interested. He knew of the earlier Belgian paleontological and geological research. He said that the mission had kept the equipment that Professor de Heinzelin had left there in 1960 and apologized that they had finally had to dispose of it a few years ago. He promised to help us find a guide and porters. He lent me his car and a driver to go to see the local chief.

The chief was located in the village of Gety. I spoke to his assistant in French but switched to Swahili with the chief, who was dressed in the dark blue Mobutu suit of the party loyal. On his lapel he had a small pin showing an upraised hand holding a torch, the symbol of the party. I explained that we were here for scientific research and were looking for fossils of the earliest inhabitants of Zaire. We had permission from the authorities in Kinshasa and I showed him the letter. He formally read the letter and after a time responded that he would authorize our travel through the Walenda-Bindi lands, which included the Sinda-Mohari area. He would have a *barua*, a letter, prepared for me and I was to come the next day to pick it up. I thanked him and departed.

Meanwhile, at the mission Dody, Bill, and Monte had been making preparations for the trek. We lightened our packs, leaving the unnecessary equipment at the mission. With the help of the missionary we began to add to the canned foods that we had bought in Bunia and made up loads for each of the porters to carry. Pere de la Faille was very concerned about the porters and that we take enough to feed them adequately. We bought from him an amazingly large amount of *mbilibo*, a mixture of corn and beans that he said the men would like. I thought that the four porters, who were all rather slight, would never eat all this food, which seemed about twice what the four of us had to eat for a ten-day reconnaissance.

After I had gotten the *barua* neatly typed in Swahili and the loads had been apportioned for equitable weight distribution, we were ready to go. For the sake of expedition solidarity I had each of the porters feel the weights of our packs, which were the same weight as their loads. I thought there might be some discussion that we were bigger and thus could carry more weight, but the men were satisfied with the fairness of the weight distribution. However, I could tell by their glances at Dody that they wondered why she wasn't going to carry a heavier load. Women in sub-Saharan Africa habitually carry much more than men could ever manage. I have seen everything from live chickens to entire beds being carried down the road balanced on women's heads or supported by a headband and resting on their backs. A baby is frequently hanging contentedly on its mother's side in a cloth sling for good measure.

We set off. This was the classic safari, I thought. On foot, with no loud engines to disturb the birds. At a walking speed so you had time to see everything around you. The people we passed were friendly and talked animatedly with our porters, whom they all knew. They wished us *njia muzuri* (a good path).

We spent two nights on our way down the rift wall, one at a mission building near the village of Maga and the next under our large tarp before the final descent to the rift floor. On the third day we reached our initial destination, the large hill known as Ongoliba.

Ongoliba lay above and about a kilometer west of the Sinda

River, a crystal-clear stream running along white sand at the bottom of the Rift Valley. It was well known as the locality at which a fossilized monkey tooth had been picked up by a Belgian zoologist in the 1950s. The tooth had been associated with a fossil fauna later estimated to be early Miocene, as old as 15 to 20 million years old. At such an age it was perhaps the oldest monkey in Africa.

I had doubts about the age of this monkey tooth. The early Miocene Epoch was a time when earth, at least the Old World, might have been called the Planet of the Apes. There was a riot of species, some as large as a gorilla, others medium-sized, and still others tiny. One species was named *Micropithecus*. By contrast, monkeys—those cousins of apes who generally have tails, cannot hang from branches, and have a different form to their molar teeth—were exceedingly rare during this time. If only one primate were to be found, as was the case at Ongoliba, and the age of the deposits was early Miocene, I was doubtful that it should be a monkey. Either the dating was wrong or the specimen had come from a different area than the rest of the fossil fauna used to date it. There was only one way to find the answer: to look.

We were the first Western scientists to make it into the Sinda-Mohari area in twenty-two years. Our foot survey took us all around Ongoliba. The exposures were impressive and unlike any that I had seen. They were almost all straight up and impossible to scale because they were sandstone that would crumble under your weight. They looked remarkably like the weathered old sandstone college spires at Oxford. At the bottom of many of the cliffs was an impenetrable growth of elephant grass, taller than you could see over. It looked like a cornfield gone mad. I thought of *Oklahoma*—"corn as high as an elephant's eye." It made surveying impossible. Even if you penetrated into the elephant grass, almost nothing could be seen on the ground surface, the whole purpose of surveying.

We did the best we could, hacking through the elephant grass with pangas, but we saw absolutely nothing. No bones. Exhausted and discouraged, we went back to our makeshift camp on the bank of the Sinda.

Over the next several days we covered a lot of territory, cutting over ridges and through valleys, always looking at the exposures. We screened sediment through a makeshift sieve in one area and found what looked like small fragments of bone. In another area far from Ongoliba I found a small fragment of fish bone. These were the only indications of fossils we were to find on this trip. I learned three years later on another reconnaissance into the Lower Semliki that we had been only three meters from a major fossil level at one point in our initial survey, but had missed it because of the vegetation covering the ground.

The exertions of the past several weeks had taken their toll on our small group. Sand from the river had gotten into our boots and rubbed the skin off our feet. When we tried to go barefoot around camp, the soles of our feet were bruised on the rocks. The insects were the worst in the evening but were never really absent. Everyone had numerous insect bites, despite the use of repellent. It was so hot that the repellent was soon washed off by sweat. One morning I got up very early to watch the sunrise, covering myself with mosquito netting that I hung from a tree limb. The netting only touched my body at my knees, and as it became light I noticed dark patches around my kneecaps. Mud, I thought. But as I moved to wipe it off, it disappeared. Two clouds of mosquitoes had been feeding on my knees through the netting.

Some of the group had diarrhea. Even the porters were beginning to grumble, despite their huge amounts of *mbilibo*, which they chowed down with relish. I overheard one of the porters saying to another while we were hacking through one particularly difficult traverse that this was "women's work." Apparently he could think of no worse description. It was time to go back to the mission and regroup.

On the way back up the rift wall our loads were lighter because almost all our food had been eaten, but the trek was grueling nonetheless. As we approached the mission house where we had spent the night on the way in, there was a group of people dancing to a finger piano, a series of metal strips of different lengths mounted on a piece of wood and played with the thumbs of both hands. They stopped when we approached, not certain how these *wazungu* might regard their nonreligious music and dancing because they knew we were from the mission. We assured them that we did not want them to stop, and we took a breather while they danced. After a while Dody, who had taken two years of African-Haitian dance in New York, began to dance with them, showing them some new steps in the process. There were screams of delight. Never had they seen a *muzungu* woman, most of whom had been nuns, dance and certainly not like this. More people came to join in, the music became faster and louder, and everybody was laughing. It remains one of my fondest memories of Zaire.

Back at the mission we decided that we would return to Bunia and charter a small plane to fly south over the rift to Goma, where we could rent a car to see at closer view some of the exposures we would spot from the air. We would then return to the Lower Semliki for another reconnaissance before our scheduled departure in about a month. We carried out this plan, but our second visit to the Lower Semliki was no more profitable in terms of fossils than it had been the first time.

## Regrouping and a Return to Zaire

On the way back to the United States, we stopped to talk to Jean de Heinzelin in Brussels about the results of our initial survey trip. He thought that we should concentrate our efforts on the logistically easier Upper Semliki region, nearer to Goma. The deposits here were not as old as the Lower Semliki, but they were fossiliferous and we could get to them. I reluctantly agreed. It was decided that the following year Jean would come with us into the field to undertake a reanalysis of the geology along with one of his former graduate students, Jacques Verniers.

Back at New York University things were falling apart. Professor John Buettner-Janusch, the chairman of the Anthropology Department and leading figure in the field, had been convicted on conspiracy charges of manufacturing hallucinogenic drugs in his laboratory. It was unbelievable. The department was bitterly divided over those who supported "B-J's" innocence and those who were convinced that he was guilty. I tried to steer a middle course because departmental politics have never been my cup of tea, but of course my failure to publicly pillory B-J, who I was not convinced was guilty, caused people to put me in his camp. When *New York* magazine came out with a cover story on "Professor Quaalude" the reputation of the NYU Anthropology Department and its graduate program plummeted.

The dean named me to a committee to find a new chairman of the department and we chose a woman who had done excellent work in sociocultural anthropology in the South Seas. But when the new chairman arrived, she was assailed by the forces in the department who opposed B-J and who had the support of a new university administration that wanted badly to put the B-J affair behind it. I was still an untenured assistant professor, and despite having had nothing to do with B-J's research or any of the legal proceedings there was concern over where my loyalties lay. Within the year the new chairman refused to sign my contract renewal. By the time I was ready to go back to Zaire as a Fulbright Senior Research Fellow in the summer of 1983 I knew that I would be unemployed when I came back to the United States.

For the 1983 field season I had received funding from Earthwatch, an innovative foundation based in Boston. The central idea of Earthwatch was for volunteers from any walk of life to sign up to work on scientific expeditions for a tax-deductible fee. After taking off some administrative costs, Earthwatch then sent the scientist a check based on the number of volunteers signing up for the expedition. Earthwatch would provide us with badly needed funds because I intended to stay in the field for an entire year as a Fulbright Fellow. Dody and a graduate student, Bill Sanders, were going with me for the entire year in Zaire. Other scientific members of the team that I had assembled were planning to come for shorter periods of time. There would be two series of three Earthwatch teams, one in the summer and one in the winter. Three Zairean graduate students, Modio, Muhaya, and Mugangu, were accompanying us for the summer fieldwork.

## THE UPPER SEMLIKI

We first drove into Ishango after dark, in our new Landrover towing a trailer of supplies and equipment. Ishango was to be our base within Virunga National Park for undertaking the survey and excavation program in the Upper Semliki Valley. The Semliki River originated at Ishango, flowing out of Lake Rutanzige (formerly Lake Edward and formerly Lake Idi Amin). We set up our tents under the Landrover's headlights with the help of some of the park guards. We slept soundly except for the occasional snorting of the hippos in the river below.

It was already getting light when I woke up. I looked out of the tent, and the sight took my breath away. Ishango had been described as "the most grandiose site on the planet" in promotional literature for the park, and I knew that Belgian King Baudouin and Queen Fabiola had chosen it for their fishing retreat during colonial times. But I had underestimated Ishango. We were camped on a high bluff overlooking the Semliki River as it flowed in a long arc out of the lake and disappeared to the north, snaking behind white cliffs in the distance. Straight ahead past the plain of the river the Western Rift wall rose up as a tremendous backdrop, now dramatically lit by the rising sun. To the south was Lake Rutanzige, a vast calm sea of blue green. I could just make out the peaks of the Virunga volcanoes south of the lake. To the north lay the snow-covered Ruwenzoris. It was spectacular terrain—but would we find any fossils?

The morning was spent unpacking and arranging supplies in one of the three buildings, now rather dilapidated, that constituted the station of Ishango in Virunga National Park. Originally built as tourist houses, complete with running water from basement cisterns, the houses' roofs now provided abodes of thousands of bats. Late in the morning Bill went to explore with Modio.

They came back around noon with a fossil that was going to occupy us for the next three months. Bill had been a quick student in class, and he had spotted a bone fragment that he thought was a piece of a human frontal bone. He was right. It was the piece of bone that constituted the middle of the bony forehead up above the bridge of the nose. In living people this area is "glabrose" (lacking hair because the eyebrows begin to either side), and thus anthropologists had given the central landmark point here the Greek term *glabella*. The breaks on the bone fragment were sharp and appeared recent, so there was a reasonable possibility that more of the skull, maybe all of it, was still in the ground.

I designated the discovery site Ishango 1. It was the top of a hill overlooking Lake Rutanzige. Not a bad place for a final resting place, I thought, but at the same time I knew that the height of the level meant that the fossil could not be too old, maybe several thousand years. The specimen was more broken than usual because buffaloes had trampled over the area. This caveat was lost for the moment, however, in the excitement of finding more pieces of the skull eroding out of the ground. We looked like a group of slightly daft pensioners engaged in an Easter egg hunt. But at the end of the day we had a plastic bag full of fossil human skull fragments ranging from about the size of a dime to a silver dollar.

Putting together such jigsaw puzzles of bone is one of the stocks-in-trade of the paleoanthropologist. Using small clues in the shape and form of the bone fragments, he or she fits together as many parts of the skull as possible. The process usually takes many weeks, if not months or years. The amount of work invested in such a reconstruction frequently means that the anthropologist develops a strong, even parental, feeling toward the specimen. For example, when his reconstruction and measurement of the brain case volume of the infamous and fragmentary Piltdown skull was challenged, the great British paleoanthropologist Sir Arthur Keith reacted by calling for one of the human skulls from the Royal College of Surgeons to be smashed into small bits and for some of the pieces to be withheld. He then reconstructed the entire skull to within a few cubic centimeters of its original form to confound his critics. The procedure of smashing a human skull for gradu-ate students to reconstruct had been a relatively common procedure in anthropology classes until recently when the costs of replacing original skulls made the practice prohibitively expensive.

The Ishango 1 skull began to take shape. I thought that the specimen would be identical to the skull anatomy of Africans living in the area today. The people indigenous to eastern Zaire have in general quite lightly built skulls, as do most Africans. But Ishango 1 showed a heavy brow ridge, thick skull bones, and a strong development of the muscular ridges on the back of the skull for the neck muscles. Despite my initial skepticism that Ishango 1 would have sufficient geological age to reveal anything about human evolution, I decided that the anatomical evidence made the specimen interesting enough to warrant excavating the site. I laid out the grid lines in meter squares oriented north-to-south. Bill was in charge of a team screening all the surface sediment within the squares, while I took the Landrover back to Goma to meet and bring back the Earthwatchers.

The Earthwatch teams had been oversubscribed. I had set the limit at twelve per team, but the interest had been so great that Earthwatch had convinced me to increase the number to fifteen. Even that limit had been exceeded, and there was a waiting list. If all fifteen Earthwatchers showed up in Goma at the appointed time—which I doubted—then we would have to put some on a small plane that I had an option to charter from Goma to the Ishango airfield. When we had first arrived in Goma I had contacted the various charter companies and missionary pilots operating out of the Goma Airport. Only one, a company named Lucas Air, was willing to try to land at Ishango. No one had used the airfield in years.

Lucas Air was run by an American, Duane Egli. Known as the "old eagle" by the Zaireans, Egli was a blue-jeaned man in cowboy boots with a shock of white hair. Like his namesake, the white-headed African fish eagle, Duane preferred gliding on the thermals over the African forests and savannas to the congested airports and bureaucratic tangles of the Western world. A man of expansive hospitality, he ran his operation in Goma like the ranch he owned in Texas near the Mexican border. His pilots bunked in the house, which was appointed with a bar above which everyone hung their hats when they were flying or out of town. We were adopted into this fraternity, stayed in the bunk house when we were in Goma, and left our field hats over the bar. Duane arranged our flights into Ishango in his reconditioned DC-3, left over from World War II, arranged cut-rate prices for fresh food in Goma to be flown up to us, and even met the Earthwatchers when they arrived. Duane never lost his amusement at the Earthwatch enterprise. He loved to tell people what I did. "He's going up there to Virunga Park and looking for bones, and he's got this whole group of American rich people—I think they're called 'earthworms' or something—and they pay him to work, so they can dig out in the hot sun."

I was surprised when all fifteen of the Earthwatchers made it to the rendezvous point in the Masques Hotel in Goma on the appointed day at the appointed hour. I was even more surprised to find that I liked them. I had expected the average Americans who might come on Earthwatch expeditions to be monolingual, overweight television addicts who would be eaten alive by hustlers as soon as they crossed the border into Zaire. But everyone had heeded my instructions in the briefing books that Earthwatch had put together, and no one had encountered any major trouble. These were resourceful, intelligent, well-traveled people of all ages and from all walks of life, and I found myself proud of the fact that they were from the same country as I was. I thought there would be a scramble for the few places on the plane to Ishango, but again to my surprise almost everybody wanted to go with me in the Landrover over what I had described as undoubtedly the worst roads that I had seen anywhere in the world. We had to draw straws, and the team members who lost went in the plane.

The next morning we set off early. The team members in the plane would fly north through the Virunga volcanoes and across Lake Rutanzige to be at Ishango within an hour. They would be met by Bill Sanders and immediately go to work on the excavation. They taunted the others by saying they would find all the hominids before we even got back to camp. The Earthwatchers going by car had a grueling two-day drive ahead of them, but they shrugged off the quips of the plane crew with the attitude that they were going to see the real Africa close-up.

Things got real soon enough. After stocking up on vegetables at a market about an hour north of Goma, we put a gerry can of gas into the Landrover. Some time later I thought I noticed that the Landrover's engine was running a little rough, but I thought nothing of it. About noon we put another gerry can of gas in. By the time that we pulled into the Rwindi Lodge where we were staying the first night, I knew that something was definitely wrong. I suspected dirty fuel. In Zaire gasoline is frequently transported by bicycle in small one-liter containers that may have contained cooking oil, kerosene, motor oil, or anything else. A fair amount of dirt may also be introduced as fuel is transferred from container to container. Because there had been no gas in the official gas stations because of a national shortage, I had been forced to buy gas in Goma from *gadhafis*, local bicycle vendors who took their nickname from the leader of oil-rich Libya. At Rwindi I took off the gas line and cleaned it out, removed the spark plugs, which were very dirty, and cleaned them off. We filled up the tank, this time filtering the gas with the finest mesh cloth we could find.

The next morning the Landrover had to climb the Kabasha Escarpment up the wall of the Western Rift Valley. We made it without difficulty, and I assumed that I had fixed the problem. I was mistaken. By the time we were into more mountains to the north and all the original unadulterated gas in the tank had been used, we began having major problems. The engine would suddenly lose power, cough, sputter, and then die. The only thing that seemed to help was cleaning the spark plugs, which I ended up doing every few miles because they were getting black carbon buildup, despite having the correct gaps. The second day we covered only about sixty kilometers and camped on the soccer field of a Catholic mission.

The next day we managed to find some gas to buy. We drained the tank and filled it up with the new gas, after carefully filtering it. But our problems persisted, and by noon we had gone only another twenty kilometers. We were now pushing the

Landrover over small hills and coasting down the inclines. The engine was running, but there was almost no power. We finally made it to a Canadian Seventh Day Adventist mission with a modern garage. I was convinced that something was wrong in the carburetor because of the fouled fuel that we had originally gotten. The garage worked on the car all afternoon but did not finish. The Canadians took us in and put all of us up in their various homes, which looked as if they had been transported into the Zairean interior from the Midwest.

The mechanic thought that the gas we had bought in Goma had been mixed with Fanta, an orange-colored soft drink about one-quarter as expensive per liter as gasoline. By pouring out about ten to fifteen percent of the gas from the one-liter plastic containers and replacing it with Fanta, an enterprising *gadhafi* could net an extra liter or two on a big sale like ours. Of course the carbon buildup from the sugar in the Fanta had an insidious effect on the engine, a fact we were discovering in all its ramifications.

The next morning, after the garage had done its best, we set off on the last leg to Ishango. The problem had not been entirely corrected, but we managed to limp in for the last ferry of the day across the Semliki to Ishango. Two old metal-framed boat hulls marked "Leopoldville 1946" had been welded together and covered with a wooden platform. A crew of four polers manned the ferry at each of its four corners. My first drive onto the Ishango ferry had not been without some trepidation, but the polers were quite professional and everything had always gone smoothly. Today was no exception and we poled first upstream, then across the river, to hit the other side at the disembarkation point. We arrived at Ishango two days late.

The Goma-Ishango drive was standard fare in Zaire. The Earthwatchers were amazed at how much patience and endurance it took to just drive to the field site. If you had any energy and resolve left when you got there, you could then start the research.

Meanwhile, much to our delight, the first team had made good on their promise. They had found another hominid. Bill

Johnson, a retired tire company executive from Georgia, had gotten up from the excavation for a break and had kicked at something protruding from the ground a short distance from the edge of the excavation grid. It was a flat bone fragment. Bill Sanders investigated, and it turned out to be a second human skull.

The Ishango hominids turned out to be much later than our main quarry, the fossil inhabitants of the Western Rift of several million years ago. While the Earthwatch teams were excavating Ishango 1 with Bill Sanders, I was surveying the older deposits for signs of the much earlier inhabitants.

Surveying in Zaire was different than in any other place that I had worked. The vegetation proved to be as difficult a problem as it had been in the Lower Semliki. We had to crawl through bushes and undergrowth along trails that had been made by hippos and bush pigs. Unfortunately the heights of these creatures were between knee and waist height for humans, which meant that in dense undergrowth we proceeded crawling or bent double. The fossils also were very sparse. I began to doubt whether we would find enough bones to make the whole enterprise worth the effort, which was immense. In the Omo we would collect in half an hour what it took all day, on a good day, to collect in Zaire. Why were we here?

## THE EXPERIMENTAL APPROACH IN PALEOANTHROPOLOGY AND THE WESTERN RIFT

Paleoanthropology has always lurched forward through a combination of fortuitous discovery and systematic follow-up analysis. In the 1890s Eugene Dubois discovered *Homo erectus* in Java not through careful hypothesis testing but because the Dutch then held the Dutch East Indies and he could obtain employment as a medical officer while pursuing his paleoanthropological investigations. Dubois agreed with Darwin's opinion that the African apes were the most closely related to humankind, and therefore he presumably would have preferred to look in Africa. When a truly early hominid was found in Africa, in 1924 by lime workers at Taung, South Africa, it was again a fortuitous discovery recognized by a young anatomy professor named Raymond Dart who would have preferred to have been at a post in England. Louis Leakey happened to have been born in British East Africa, and all the discoveries made by him and his family are within 500 miles of his birthplace. The pattern of fortuity has continued into the present day. The fabulously rich site at Afar, Ethiopia, that yielded the Lucy fossils named by Donald Johanson *Australopithecus afarensis*, had been found by a French geologist interested in plate tectonics. Johanson had gotten involved because his girlfriend at the time was French, and he frequently ate dinner with the French team in the Omo. The important site at Laetoli, which preserved the first evidence of hominid bipedalism, was found when paleontologists were having an elephant dung fight. A dried piece of elephant dung scored a direct hit on one combatant, knocking his glasses off. As he groped on the ground for them, he recognized for the first time the fossilized footprints of animals in the petrified volcanic rock.

I felt that with a more deliberate experimental approach to the field aspects of research would yield important results. Rather than limiting oneself to the vagaries of where one had been born, what language one spoke, what field site had the most comfortable climate to work in, or to any number of other irrelevant factors, I proposed to undertake fieldwork in the area optimally situated in space and time to test the hypothesis of the ape-hominid split. The Western African Rift is that place.

We finished the Ishango 1 excavation and the Earthwatchers departed. We concentrated on intensive surveying in the older reddish, ironstone-encrusted deposits known as the Lusso Beds. We began to build an inventory of identifiable fossils.

The fossils were old. They matched the species known from the well-dated sites of East Africa at over 2 million years old. This age was not surprising because there were related, though still poorly known, sites in Uganda about this same age. But I was amazed that over half of the fossils that we were collecting were antelopes—antelopes with high-crowned molars. Antelopes were denizens primarily of the African savannas, and I had hypothesized that the fauna we would discover at earlier time periods in the Western Rift would be forest-adapted, based on our proximity to the present-day African forests and on some forest species preserved at the old Uganda sites. Some antelopes were leaf-eaters or browsers and some were grass-eaters or grazers. The grazers had molar teeth with higher crowns because they had adapted to chewing blades of grass with their hard silica phytoliths, which wore down their teeth as they chewed. The conclusion was inescapable. We had discovered a new savanna environment in Central Africa at over 2 million years ago immediately adjacent to the Central Forest Refuge. This finding meant that, at least at this time period, Central Africa was not going to tell us anything about the ancient forests with their ape ancestors, ancestors that we so badly wanted to find. Central Africa had become by two million years ago an ecological outpost of the East African savanna biome. We certainly were not going to find any apes here, but could we find their bipedal relatives?

The first clues had been found years before. In 1954 Belgian geologist Jean de Heinzelin discovered three stone flakes in a test trench dug by his workers in the Lusso Beds at Kanyatsi on the north coast of Lake Rutanzige. The flakes had the characteristic form of stone tools struck off a larger core stone and resembled tools made by early hominids at Olduvai Gorge and other early sites. Subsequent detailed excavation failed to turn up any more artifacts so de Heinzelin was forced to discount them as incontrovertible evidence of early hominid presence in the Western Rift. I invited Jean to come back to Zaire twenty-two years after he had left to pick up the trail.

Although Jean de Heinzelin de Braucourt is a French chevalier and has an ancestral estate near Bordeaux, he is a scientist and empiricist to the core and is egalitarian to a fault. In the field he first takes his shirt off, turns lobster red for a day or two, and then gradually browns. His razor-thin white beard follows his jawline around to his chin and sets his tanned face off in strong contrast. An outgoing personality, he talks easily to young and old alike, but he can quickly turn gruff, adopting either an aloof aristocratic or, more frequently, a laissez-faire bohemian attitude. The latter characteristic made most people like Jean, but the former made them keep their distance. Women particularly liked Jean.

Jean and I first became friends in the Omo when he started eating breakfast with me before everyone else, at 5:30 A.M. I had discovered that the flies did not become active until about 5:45 A.M., thus allowing one to have a pleasant breakfast uninterrupted by the buzzing that was ubiquitous the rest of the day. Jean began to eat American cornflakes for breakfast, a new experience for him. One morning toward the end of the field season I discovered that the last cartons of UHT milk had all sprung leaks and had curdled, so I guessed that we would have no more cornflakes. Jean didn't hesitate. He poured the carton onto his heaping bowl of cornflakes and said, "Cornflakes with yoghurt. Excellent."

When Jean came back to Zaire it was quite an occasion. All the schoolchildren in Zaire learn in their history classes about his discovery at Ishango of the earliest Zaireans, *l'homme d'Ishango* or Ishango Man. The Ishango population had subsisted on abundant fish and had fashioned elaborate bone harpoons 20,000 years ago. They had developed a system of counting that is among the earliest known anywhere in the world. As we drove past the little villages, entire schools turned out with their teachers to line the road and cheer de Heinzelin. Jean said he felt like Belgian King Badouin. Marveling at how anyone could have known we were coming at that particular moment, I was reminded of the missionary's statement at Gety, "there are no secrets in Zaire." We stopped for the night with our Canadian missionary friends, but we could not leave the next morning until Jean had addressed the assembled mission school.

Working with Jean in Zaire was different from our previous collaborations in Ethiopia and Libya. Jean had started his career in Zaire when it was the Belgian Congo. He initially had made his scientific reputation on the discoveries and publications from the Upper Semliki, including Ishango. We were here to extend his discoveries of over two decades before, but I sensed a feeling that we were almost invading a part of his past.

This place existed in de Heinzelin's memory as part of the Belgian Congo, a world now vanished and most of its inhabitants ghosts.

After a quick reconnaissance of the area around Ishango, to which we were going to return for more in-depth work later, we set off to the Lower Semliki to review our initial field observations the previous year. It was imperative to start before the rains began in earnest if we were to get in and out again. We never made it. The rains started early and along the roads truck after truck was mired, wallowing and listing like injured elephants. The smaller vehicles, cars, pickups, and Landrovers, squeaked by on makeshift tracks hacked out of the bush. This was life in modern Zaire and I was already used to it. In the days of the Belgian Congo, however, the colonial administration boasted that a car could drive anywhere in this vast country with a full glass of water on the dashboard at a consistent speed of fifty kilometers an hour without spilling a drop. Jean inexplicably became increasingly nervous as we drove and encountered each new obstacle. He wanted to turn back. He finally explained that while driving this road to escape his captors in 1960 pygmies from the forest, a group that seems to inspire universal fear in Zaire, began felling trees across the road in order to commandeer cars. He had last driven this route in a cold sweat, not knowing which bend in the road was going to be his last. The trucks mired every few kilometers were a bit too close to a recurring nightmare. We came upon two tractor-trailers wedged together in a sea of mud, flanked by a hill on one side of the road and a ravine on the other. There was no way to get through. We turned around and went back to Ishango.

Our goal now was to build up as much data as we could on the Lusso Bed fauna, archaeology, and geology around Ishango before Jean had to return to Belgium. We went back to Kanyatsi and discovered a number of stone flakes on the surface of the deposits, now exposed by erosion after so many years. Looking at the flakes with a hand lens we could see that they still had bits of ironstone adhering to them. Its presence meant that these artifacts had almost certainly come from the

Lusso Beds and thus were as old as the flakes originally dug up by the workers in 1954.

To nail down the argument for the presence of early hominids in the Western Rift we needed a detailed excavation with artifacts in place, or in situ. I wrote to Jack Harris, an archaeologist then at the University of Wisconsin, Milwaukee, who had been a graduate student with me at Berkeley, and asked if he could undertake the challenge. He agreed and came to Zaire the next year.

## FOSSILS AND THE EARLIEST STONE TOOLS

In 1984 Jack Harris decided that the best place to prove the question of whether stone tools were present in the Lusso Beds was not at Kanyatsi but at a site called Senga 5 that I had found the previous year about five kilometers up the river. He had seen a number of small stone flakes on the surface here and thought that there would be a good chance of discovering more underground. He set up a meter-by-meter grid system and began a systematic excavation that was to last three years and move tons of sediment. The site was designated Senga 5A to distinguish it from the larger collecting locality.

Senga 5A turned out to be the richest site we found in the Lusso Beds. Not only were there quartz-flake tools in profusion, some with delicate razor-sharp edges through which you could see light, but there were associated fossils. The early hominids that had made the tools had shared the area with extinct gelada baboons (*Theropithecus*), giant pigs (*Notochoerus*), and a variety of antelopes, rhino, and giraffe. One fossil was of particular note. It was the bottom shell, the plastron, of a large tortoise. Our expert on turtles, Peter Meylan, thought the specimen probably represented a new species, a sort of African Galapagos tortoise. Even more interesting from an anthropocentric standpoint were the incised marks that Meylan noticed on the side of the plastron in front of the back leg. He had noticed that turtle hunters today frequently cut the skin against the shell at this point to get into the shell and skin a tortoise. He was convinced that the incised marks on the Senga 5A specimen were some of the earliest evidence for meat-eating in hominids. In the lab the cut marks were subjected to a battery of microscopic examinations and comparisons that confirmed that the cuts likely had been made by hominids using stone tools and not caused by bite marks by carnivores, scratches by sand grains, trample marks by an animal's hoof, or accidental damage by an excavator's probe. Jack's excavation uncovered some 500 stone tools associated with Lusso Bed fossil animals, a dramatic demonstration that stone tools were present in the Lusso Beds. But how old were those beds?

The Upper Semliki did not seem to have any tuffs, volcanic ashes rich in potassium that had proved invaluable in potassium-argon dating of the fossil sediments in East Africa. The geologists were puzzled by this absence, because we were in a rift valley, which by definition is bounded by large faults. Such large faults are frequently associated with volcanoes, which are simply cracks in the earth's crust through which magma and associated gas and ash escape. The cracks were there, but the volcanoes, at least in Lusso times, were not. The Virunga volcanoes, which can be seen from Ishango on a clear day, are active today, but they produce lava, not ash, when they erupt. Possibly the volcanoes in Lusso times, if there were any, were similar and produced only localized sheets of lava. If they did so, however, we certainly did not find any indication.

Without potassium-argon dates we had to rely on a comparative study of the animal fossils, or biostratigraphy. The paleontological survey recovering those fossils proceeded in tandem with the archaeological excavations.

My survey teams covered the entire area of outcrops of the Lusso Beds in the Upper Semliki; they crawled through hippo trails, walked along the river banks and lake shores, and trekked over long expanses of savanna. The area is about thirty kilometers along the river and ten kilometers along the lake, extending a kilometer or two inland in both cases. A National Geographic Society grant had bought us a small outboard motor boat because a number of sites along the water's edge could not be seen or easily reached from the landward side.

With the boat we could also survey the western bank of the river. We would locate sites on the river or lake from the boat, land and make collections, and record the location on our aerial photographs. We would then return the next day by Landrover and on foot, hacking our way through the bush, to check on the accuracy of our aerial photo plotting and to cover the site once more for fossils.

The voyages in the boat were not for the faint-hearted. De Heinzelin told me in no uncertain terms that I was insane to take a boat on the Semliki River. He had nearly been killed when a bull hippo attacked his wooden pirogue in the mouth of the river, collapsing the side and sinking the boat. A hippo had attacked our Canadian missionary friend while he was swimming at Ishango, biting him through the upper thigh and buttocks and barely missing the femoral artery and sciatic nerve. British doctors across the border in Uganda had sewn him up. There were also rapids on the Semliki, mainly at Senga, and in some places the river was shallow-bottomed. You could be capsized in the rapids or lose a propeller on a rock on the bottom.

I do not believe in taking unnecessary chances. I fully agree with Sir Vivian Fuchs, the British explorer who, when asked to recount his "adventures" in traversing Africa from the Cape to Cairo and in reaching the South Pole, said simply that he hadn't had any. "Adventures," he is reported to have said, "are for fools."

However, there was no other way to locate expeditiously all the fossil-bearing sites in the Upper Semliki without using the boat. So I initially made several trial runs near camp among the hippos. I found that they are remarkably like cows, around whom I had learned to negotiate as a boy in Virginia. They can be herded in the water if you cut them off or if you stay behind them, but you must remember that when alarmed they want to stay in the river. And you can never run over them because when alarmed they rear up out of the water and can easily overturn a small boat. The forty-horsepower outboard allowed us to move faster than the hippos and thus outmaneuver them. De Heinzelin's ponderous pirogue probably had run over and enraged a big bull, who had reacted predictably. When I went surveying I always had a spotter in the front of the boat who kept track of the hippos as they submerged. Following park rules, I also took a guard, but a shot from one of their .22 rifles would hardly have deflected an attack by a charging hippo. In several years of surveying by boat we never had an incident with the hippos on the river.

The only serious set of rapids in the Upper Semliki, at Senga, were caused by an underwater outcropping of an iron-stone level of the Lusso Beds. When I had reached Senga in my surveying, I decided to run the rapids. I knew there would be no problem in descending, but returning against the current would be tricky. We had had a great day of discoveries north of Senga because we were in virgin territory. On the return I knew that we had to be careful or all of the fossils might be lost over-board if we capsized. As we approached the rapids I landed on the river bank and unloaded my three crew members with the fossils just to be on the safe side. I then started up the rapids, keeping to the open water. The outboard was at full throttle and just barely made progress against the current. The pro-peller hit a rock on one or two turns but eventually I was able to negotiate a diagonal path through the rapids.

The fossil collections began to build up, but very slowly. Within four years we collected some fifty-one species of verte-brate fossils. By matching the species of animals and their stage of evolution, as evidenced by their tooth form and anatomy, it was possible to match our fossil site with those in East Africa, which are well dated. The Zairean sites came out to be between 2.0 and 2.3 million years old. This made the Zairean stone tools as old as those from Omo and as old as any in the world.

The confirmation of stone tools, and at such an early date, from the Upper Semliki met with great interest. First of all, these were the first very early stone tools confirmed from cen-tral Africa and the Western Rift. Second, they were found in an environment generally thought to have been at least partially forested, although our continuing research was modifying this idea. And third, the inception of stone tool making was a momentous landmark in the human evolutionary career, which is still poorly understood. What preceded it? Were the tool makers uniquely *Homo habilis*, or could the coeval robust australopithecines have had a hand in the production of tools? The Upper Semliki fossil deposits promised some answers.

But something was strange about the geology at Senga 5A. The geologists were not able to make a number of their observations fit with the rest of the Lusso Beds. The dip of the beds at Senga 5A differed from the beds in the surrounding area, and there were some calcareous nodules that did not normally occur elsewhere in the Lusso Beds. Something was wrong.

Nothing is more worrisome to paleoanthropologists, and especially to archaeologists, whose data are more context-related than are the fossil bones studied by paleontologists, than a problem with the geological interpretation of a site. Everyone remembered Louis Leakey's first discovery of a fossil human mandible in the early 1930s at Kanam, Kenya, He claimed that it was found in beds of early Pleistocene age. His consequent claim of great antiquity for the human species, however, was met with skepticism and then doubt as later geological work could not confirm the exact stratigraphic location of the find. Leakey said that heavy rains had caused slumping and obscured the original spot of discovery. It was rumored that Leakey never made it into Great Britain's Royal Society, despite many decades of contributions, because of that misstep.

Jack Harris became concerned about the situation. We went back to the field in 1988 in full force. All the geological team was there: Jean de Heinzelin and his former student Jacques Verniers, Peter Williamson from Harvard and his graduate student Paul Morris, and David Helgren, a geomorphologist from the Monterey Institute in California who was working mainly with archaeologist Alison Brooks. A flock of graduate students from Rutgers, Harvard, Wisconsin, and George Washington were there on the excavation team. Senga 5A was subjected to an excruciating battery of analyses and tests designed to figure out whether or not the site was as old as we had claimed.

De Heinzelin was the main nay-sayer. One day he took me to the site by myself, away from the chorus of voices, and methodically showed me his evidence. There were small inclusions of white nodules of calcium carbonate, which did not occur anywhere else in the Lusso Beds. He knew. He had walked all of them, beginning thirty years before. Then, the dip of the site was at a different angle than the surrounding Lusso Beds. And he could not find continuous levels of the typical Lusso Bed ironstone; there seemed to be large chunks of it, but it petered out into the surrounding sand. He thought that there was only one conclusion. The Senga 5A site was not an in situ deposit of 2-million-year-old fossils and artifacts; instead, the finds had been eroded and redeposited by the Semliki River in the late Pleistocene, only a few tens of thousands of years ago.

Some wanted to fight the inevitable conclusion, but I was not surprised when the results came in from geochemical analyses on the Senga 5A sediment from Peter Williamson's lab. The sediment contained a high level of the mineral perovskite. No other Lusso Bed sediments contained this mineral of volcanic origin that was found only at the top of our sedimentary sequence, in rocks not older than the latest Pleistocene. De Heinzelin had been right.

Jack had the demeanor of a man headed for the gallows as he began working on the paper reporting our findings. But as we put the observations together anew, another fact was striking. Although there was redeposition of the Lusso sediments by the river in late Pleistocene times, all of the fossils were of the heavily mineralized, iron-stained Lusso variety. Their identification again confirmed the circa 2-million-year-old age. A number of the quartz artifacts also were iron-stained or had red Lusso ironstone adhering to them. Our conclusion was that Senga 5A had indeed been redeposited but that it was a re-covered fossil site of Lusso age. In other words, there was a basic integrity to the fossil bones and artifacts even if Jack and the archaeologists could not use the spacing of the tools to deduce how early hominids may have dropped the tools on the landscape. That information had been lost when the site was redeposited.

In our opinion, there was not a serious question about the age of the Senga 5A stone tools, but we wanted to discover a clearly in situ occurrence to put the matter to rest. Jack put in a small excavation at Kanyatsi, de Heinzelin's original artifact site, where I had first suggested that Jack excavate. A single quartz flake was discovered in situ at Kanyatsi, confirming de Heinzelin's first discovery in the 1950s.

## SENGA 13B

I wanted to find where the fossils and artifacts had come from at Senga 5A if they had not been in situ. While Jack was excavating at Kanyatsi, I started an excavation at a site two kilometers south of Senga 5A, called Senga 13B. In 1983–1984 I had found there some limb bones of an elephant eroding out of the sediments and had collected some other fossils. Jacques Verniers thought that the ironstone level at Senga 13B was probably the same level from which the Senga 5A bone and stone tool assemblage had come. I reasoned that we should be able to find in situ artifacts with fossil bone there if anywhere in the Lusso Beds.

John Gatesy, then a graduate student at Yale; my former UCLA and NYU graduate student Paris Pavlakis, now at the University of Ioannina in Greece; and Meleisa McDonell, a former podiatric surgeon, graduate student at Johns Hopkins University, and eventually my wife, all collaborated with me over four years of excavation of Senga 13B. We completed over forty square meters of excavated area at Senga 13B and recovered the largest assemblage of fossils from an excavated site in the Western Rift.

Soon after beginning the Senga 13B excavation, we began to find the rest of the elephant. It had been a juvenile because the ends of its long bones had not fused. But, strangely, the bones were cracked, as if other elephants had stepped on them. De Heinzelin's view of the site conformed with that idea. The sediments showed that there had been a lot of surface disturbance, probably trampling by animals, of the top layer. Some bones had been pushed several tens of centimeters into what was then mud. We finished excavating and mapping out the elephant level. We did not find fossils of any other animal or any artifacts. Clearly this level could not have been the origin of our diverse fossil and artifact level at Senga 5A. It had most likely been a bog or swampy area next to the lake when it was deposited. A young elephant had died or been killed at the lake margin, and his kin had trampled his bones into the mud as they had come to drink or feed. There was no evidence that hominids had found or had eaten any of the elephant.

We dug deeper, down to a lower ironstone level at Senga 13B. Here we found many fish fossils. But in among the fish were crocodile teeth, turtle bones, and mammals. The sediment here was coarser, sand, where the elephant level had been clay. All the bones and teeth were small and broken. On the average we found one to two mammal teeth per day, not a bad yield because this was about the same rate of discovery as I had had during surface survey.

We continued the Senga 13B excavation for three years, until 1990, the last year we were able to work in Zaire. Over 1,300 fossils were excavated, mapped, and identified. The site told us a lot about the environment of the Upper Semliki over 2 million years ago, but we did not turn up any fossil hominids or any definitive evidence of their stone tools. A few pieces of angular quartz were in the excavation, but Jack was not convinced that these were "from the hand of man," in his words. We had not demonstrated that the Senga 13B ironstone level had the same composition of fossils and artifacts and could thereby account for the Senga 5A deposit.

FUTURE RESEARCH

After eight years of effort, I had to assess whether the Zairean project had paid off and whether it was worth pursuing any further. By most traditional anthropological standards the Semliki Research Expedition had been a failure, if the yardstick of numbers of hominid fossils were used to assess it. If you added that working in Zaire was far more difficult, time-consuming, and costly than any other place in Africa that I had encountered, then the cost–benefit ratio left no doubt as to future work in the Western Rift. Most of my peers apparently felt this way. Our grant proposals to the National Science Foundation to return to the field in Zaire were turned down.

I was far from convinced that the Western Rift had been mined out. I admit that I gave short shrift to the opinions of my colleagues. Most of them are lab people who consider field-work too risky, even if it is successful. They would rather study specimens that have been collected by the few hardy souls who are willing to venture out to find them. After all, they reason, unproductive search time and grant proposal writing do not lead to publication, and this is many peoples' yardstick of productivity, if not personal worth. I knew that most of my reviewers had very little idea of the conditions in the field, and, from their armchair assessments, prospects probably did not look bright for paleoanthropological research in the Western Rift.

The Western Rift of Africa is probably the most important place that paleoanthropologists could now be working. I conceive of the Western Rift as an edge, and edges are always interesting places—cliffs, seashores, islands (which are just circular edges). Here topographies change. The animal and plant species that live on either side of the edge come into contact. They sometimes interact, and how they interact can be fascinating. The Western Rift is the edge, the dividing line, between humankind and its nearest living relatives, the chimpanzee and the gorilla. But why is it the dividing line? Is it forest versus savanna? Is it mountain versus lowland? Why then does the molecular evidence now indicate that humans and the forest-living chimps share a common line after the split of the gorilla? The story is complicated, and we will never know the answer unless we investigate the area where the splits occurred. Right now the African Western Rift Valley is a good candidate for where these splits might have occurred.

The Western Rift is also an excellent laboratory for investigating the history of environmental change, and hominids' responses to those changes. The fossil deposits of the Lower and Upper Semliki Valleys in Zaire, and in various parts of western Uganda, preserve a patchwork of different times and different habitats. These deposits record when forests expanded out of central Africa and into eastern Africa and when savannas extended from eastern Africa into central Africa. This shifting pattern of forest and savanna, mediated by global climatic changes as well as by local topographic changes, undoubtedly had a huge effect on the evolutionary history of our closest living relatives, the forest-living apes, as well as our own lineage. Stone artifacts seem to occur in only the most arid stratigraphic levels in the Upper Semliki. This finding implies that hominids were indeed denizens of the open savannas. But much more work needs to be undertaken to ascertain the relationships of the various fossil sites in the Western Rift and to test the scenarios that we have been able to postulate.

My expedition is now preparing for new research on the Ugandan side of the Western Rift Valley in deposits that have yielded indications of forested habitats back to 3.5 million years old and possibly older. This will be a new chapter of fieldwork in central Africa and will build on what we have learned in Zaire. For now, though, more specific answers to the question of the "western missing link" must remain unknown.

# 8

# Paddling Upstream in the Cerebral Rubicon

By now it should be obvious that the search for the missing link is not easy. It is certainly not over, nor will it be concluded soon. I have argued that hominids split from chimps in the west of Africa, around 8 million years ago, but I can't prove it. Others argue that the split occurred in the east, but they can't prove it either.

What I think I have demonstrated, however, is that our understanding has improved tremendously in the last forty years. Whether or not the Western Rift is essential, whether or not Lucy is part of a different species than *Australopithecus afarensis*, whether or not North Africa played an important part in early hominid evolution, we can at least be relatively certain of a number of specifics:

- Hominids split from gorillas first, at around 12 million years ago
- Hominids split from chimps last, at around 8 million years ago
- The two lines of australopithecines split prior to 2.5 million years ago and maybe as early as 5 million years ago

- The genus *Homo* and the first stone tools appear at 2.5 million years ago

We know these specifics largely because of the techniques developed in the wake of the Cold Spring Harbor conference in 1950. Because we study the entire context of evolution, using techniques and perspectives from geophysical dating to ethology, and from molecular genetics to population biology, we have come a long way from the ideas of the missing link of the 1950s. In the final chapters I will discuss the progress we have made in two particular post–Cold Spring Harbor fields: the study of tools and of climate.

## TOOLS AND THE EVOLUTION OF THE HUMAN BRAIN

No other trait has been more important in evolutionary debates than the size of the brain. Long before Darwin, anatomists had noticed that virtually all animals, even the relatively brainy primates, have significantly smaller brains than do people. And the animals that do have larger brains, such as elephants, have huge body sizes, so their brains are not relatively larger. When in human evolution did big brains first come into being? And why?

Theories have abounded. In the early phases of anthropology, when fossil data were rare or nonexistent, debate centered around the comparative anatomy of the living primates. Thomas Henry Huxley, the champion of Darwin's theory of natural selection in England, studied the brains of apes and of humans. He concluded that there was nothing structurally unique in the human brain. It was simply an enlarged version of the chimpanzee brain. St. George Mivart disagreed, maintaining that there were unique structures in the human brain. Huxley's and Mivart's running debate on the presence or absence of an obscure part of cerebrum, the *hippocampus major*, seems at first to be an absurdly esoteric exercise. However, on its outcome turned the issue of whether the human brain could be derived from an animal precursor or whether it was an evolutionary novelty, so different from similar structures in closely related organisms that other hypotheses would be needed to explain its origins. Huxley's views prevailed, but this acceptance of anatomical reality did not spell the end of the idea that the human brain is something that transcends nature. Many scientists believed that the human brain was the first and most important change that set hominids off on their evolutionary trajectory away from the apes. Some, such as Sir Grafton Elliot Smith, even believed that the brain in some way directed its own evolution.

Piltdown Man, a large-brained fossil discovered first in 1908 in very early Pliocene-aged gravels in southern England, provided the greatest support for the hypothesis of the primacy of the big brain in hominid evolution. Unfortunately, both for science and for the scientists who had pinned their careers on it, Piltdown turned out to be fraudulent.

One of the greatest successes of modern paleoanthropology, by contrast, is its demonstration of the nature and rate of brain size *increase* over the span of the last 3 million years—the opposite of the Piltdown idea. Instead, it is argued, brains evolved along with their hominid owners. Brains are not somehow magically large in all hominids. In fact, long before 1957, many paleoanthropologists had already rejected Piltdown as an aberrant amalgamation of human and ape fossils and had discounted it as an important part of the documentation of human evolution. The tide of discovery had turned to Africa, and beginning in the 1920s increasing numbers of australopithecines were discovered in cave sites in South Africa. These hominids had human-like teeth, but they had quite small brains, only slightly larger than apes. In these two traits then they showed exactly the opposite trend from Piltdown. Australopithecines showed that hominids had not sprung large-brained from the head of Zeus.

If there was a gradual increase in brain size through time and if large brain size defined hominids, where should one draw the line between hominids and apes? The australopithecines forced the question. One answer, proposed by England's Sir Arthur Keith, was the "Cerebral Rubicon," an amount of brain volume that defined human status. Keith suggested its value to be 800 cubic centimeters, about the size of a small cantaloupe or a large grapefruit. Above that value, a brain would be human; below it, something else. The Rubicon was a river that Julius Caesar crossed in his conquest of Gaul, beyond which he could not return to Rome. Keith used it as a metaphor for crossing the threshold to humanity, but the Cerebral Rubicon has meant troubled waters for anthropology.

First of all the Cerebral Rubicon is a typological concept, an ideal condition that allows no individual variation. In a population of early hominids that had a mean brain value of 850 cubic centimeters, for example, statistically a good proportion of individuals would fall below the magic 800 mark. Were they not hominids, even though they were fully part of a population that was? The Cerebral Rubicon is a concept born before the modern biological concept of populations and of statistical variability.

Second, the Cerebral Rubicon concept suffers from that malady shared by any hypothesis that attempts to define an entire species on the basis of one characteristic alone. To be defined as accurately as possible, species need to be identified and named by taking into account as many biologically meaningful characteristics as possible. Even Keith would not call a large male gorilla a hominid if its cranial capacity happened to exceed 800 cubic centimeters. The gorilla has many other characteristics that clearly make it a gorilla. Thus, most anthropologists today realize that cranial capacity, or any single trait alone, should not be used to distinguish hominids from their closely related primate relatives and fossil ancestors.

As anthropologists moved away from arbitrary and typological concepts like the Cerebral Rubicon and as bona fide fossil evidence of early hominids began to accumulate, it became apparent that brain increase had occurred significantly later than the adoption of erect posture, bipedalism. What led to the phenomenal increase in brain size and intelligence in hominids? Anthropologists have been grappling with the problem for well over a century.

Darwin hypothesized in 1867 that upright posture had led to a freeing of the hands, which in turn had allowed hominids to wield weapons for protection. With the defensive functions of the large canine teeth lost, natural selection acted to reduce their size. Hominids lost their fangs.

Anthropologists then took the ball and ran with it. If weapons and tools had become the primary means of adaptation in hominids, then this was culture, that learned system of behavior unique to human beings. To be culture-bearing, hominids had to be intelligent and therefore large-brained. The connection between tools and large brain size was thus established.

When Louis Leakey first went to Olduvai Gorge in 1931 he had made a wager with the German geologist Hans Reck, who had investigated Olduvai before World War I, that he would find stone tools there as soon as they arrived. In this he was successful, picking up several stone choppers and hand axes scattered on the ground. Once hominids began making stone tools, they made lots of them, and because the small chipped stones are virtually indestructible, they are easily preserved in fossil sites. Finding the makers of the stone tools at Olduvai took substantially longer. Not until 1959 did a skull of a robust australopithecine, named *Zinjanthropus boisei*, turn up. Leakey immediately anointed him as the maker of the stone tools, but Zinj's heyday as an Olduvai culture-bearer was short-lived. In 1961 another hominid, this one large-brained and human-like, was discovered. Leakey named it *Homo habilis*, "handy man" in Latin, referring to this hominid's presumed ability to make and use the stone tools that had been turning up at Olduvai for decades. *Homo habilis* then reinforced the connection between large brain size and stone tool making. Zinj, now known as *Australopithecus boisei*, was conceived of as sort of a hominid gorilla, dull-witted and slow, who could not make tools.

But there was a problem with this scenario. If Darwin was right and small canines were related to weapon and tool use, why did robust australopithecines have small canines? Alternatively, if robust australopithecines had been tool users why didn't they have large brains like *Homo habilis*?

There are a number of potential answers to these questions, none of them entirely satisfactory. My own explanation relates to that theoretical construct, culture. At the level of *Homo habilis* and the robust australopithecines, we have what we can term low culture, *bas culture*, to distinguish it from the *haute culture* of modern people. *Bas culture* did not evolve like modern culture does; there was not a new style of stone tool every year like Paris dress styles or Detroit auto models. Two systems of *bas culture* could exist and probably did. Both hominids probably used tools to some extent. To what extent we do not yet know. We might guess that *Homo habilis*, because of its larger brain, may have been ahead of the robust australopithecines in use of tools. The robusts, on the other hand, could eat a wider range of tough foods because of their formidable molars and great crushing jaws.

Another problem with Darwin's original formulation presented itself when hominids with reduced canine teeth, such as *Australopithecus africanus* and *afarensis*, were found in fossil deposits that lacked all evidence of stone tools. Did these hominids use a type of tool that did not become preserved in fossil deposits, or was the hypothesis incorrect?

Anthropologist Cliff Jolly thought that Darwin's hypothesis was incorrect. He suggested that small canines had evolved in response to changes in diet. Jolly studied the teeth of the gelada baboon, which has smaller canines than the much more common savanna baboon. He suggested that geladas' canines had become reduced through evolution because they had to chew side-to-side with strong crushing movements. Their diet is composed of tough, small seeds and nuts. Jolly thought that hominids had undergone the same pressures of natural selection and had evolved their small canine teeth because they had switched to a small-object savanna diet that was tough to chew. Jolly's hypothesis became known as the "seed-eater hypothesis" and found favor with anthropologists who were grappling with the problem of reduction in canine size in early hominids before the appearance of stone tools.

But as with all analogies, in art or science, Jolly's hypothesis was not a perfect match for early hominids. Other animals, such as warthogs, have significant crushing ability in chewing, yet have huge canine tusks used for defense. And in the human chewing cycle our jaws do not extend to the side to the extent that the movement would be significantly hampered even if we had long canines. Long canines clearly would not help us in eating, but I have never believed that selection for chewing efficiency alone would be sufficient to reduce the canine teeth of our prehominid forbears. The warthog example shows that if selection acts to increase crushing ability in chewing but large canines are still needed for defense, both capabilities can be retained. I think that in hominids there had to be relaxation of selection for large canines, which are so important in ape defensive displays and fighting. I am consequently not so quick to throw out Darwin's old idea that tool use played an important part in initial hominid evolution.

Could there have been a time when hominids used tools as a regular and essential part of their adaptation to life in the savanna, yet had not discovered the use of stone as a raw material? I thought the answer was yes.

Raymond Dart's concept of "osteodontokeratic" tools, or non-stone-tools, had been around for years. Paleoanthropologists mentioned them in their papers, books, and classes, more out of deference to Dart than due to any belief in their authenticity. But in 1986 C. K. Brain of the Transvaal Museum published evidence that there were indeed alterations to bones found at the cave site of Swartkrans that could not be explained as effects of chewing and breakage by hyenas and other carnivores. Brain believed that these altered bones were tools used by pre-stone-using australopithecines.

Another discovery from South Africa tended to reaffirm the antiquity of tool-using ability in early hominids. Randall Susman of the State University of New York at Stony Brook found that the thumb bones of *Australopithecus robustus* show the same ability to rotate for a precision grip that modern humans have. Apes tend to use the sides of their thumbs when they pick up something delicate, whereas people turn their thumbs in to meet the other fingertips. This ability to fully oppose the thumb had also been shown in other stone-tool-using hominids in the human lineage, particularly *Homo habilis* from Olduvai Gorge (based on the Olduvai Hand, specimen O.H. 8). The further discovery of this trait in robust

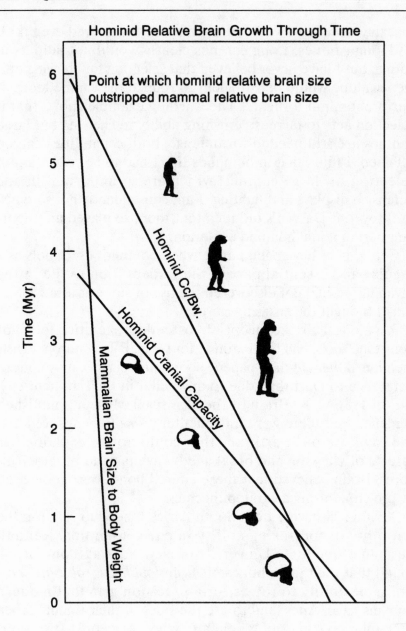

Hominid Relative Brain Growth Through Time

australopithecines showed that the adaptation was likely an ancient one, predating the split between *Homo* and *Australopithecus*. Finally, studies by Mary Marzke at the University of Arizona showed that the hand bones of the even earlier *Australopithecus afarensis* show fine manipulative ability long before stone tool use came onto the scene. I conclude from the anatomical data that the earliest known hominids could make and use tools of some sophistication—that is, more complicated than a chimp termite-fishing probe. These tools were certainly mostly of wood or bone and thus much less likely to be preserved as archaeological artifacts.

Tool-making ability traditionally has been thought of by anthropologists as the *sine qua non* of humanity. "Tools makyth man," Kenneth Oakley of the British Museum intoned years ago. Does not suggesting that the earliest hominids made tools imply that they had a level of intelligence usually associated with later stone-tool-using hominids? The answer to this question is definitely yes, but testing it is more difficult.

In one of my studies on the Omo hominids that I published in 1985, I looked at the trend in brain size growth in hominids over time. Although it is not a perfect indicator of intelligence, brain size compared to body size is a fairly good predictor of what we generally term intelligence—the ability to take in information from the environment, assimilate it, remember it, and make decisions upon it. I plotted hominid brain-to-body size through time on the same graph as brain-to-body size of mammals (see chart). The slope of the hominid line was much higher than the mammal line, indicating that the hominid brain had increased in size through time at a much higher rate than mammals, an expected conclusion. It also showed that as hominid body size increased, brain size increased proportionately more than in mammals. The conclusion is thus that hominids are genetically wired to have bigger brains than the average mammals, including apes.

What I found most intriguing about this comparative graphing of hominid brain size growth through time was that the hominid line intersected the mammal line at a time corresponding to 6 to 7 million years ago, within the time span for the chimp-human split. The line had to be extrapolated over a big area of blank graph paper, because there are no fossil hominid skulls with body size estimates before 3.2 million years ago. But if this speculation is true, hominid brain growth, also known as encephalization, was one of the first adaptations that characterized the hominid lineage. What could have been causally related to this brain growth? I thought the answer was likely tool use and the food-getting adaptation on the savanna that relied on it. There might even be a place for Darwin's idea that weapon use may have played a role.

## A Scenario for Hominid Brain Evolution

The evolution of the brain in the earliest hominids can be hypothesized as follows. Small-bodied, imperfectly bipedal hominid-chimps became isolated from their chimp-hominid confreres living in the forests and on the forest fringes. Increasing patchiness of forests and woodlands put significant strains on both populations. Body size increased in response to predator pressure in the chimp-hominids that were to evolve into the chimps. The proto-chimps reaped other ecological advantages by having bigger body sizes, as discussed in the previous chapter. But the chimps retained the ancestral brain adaptations that characterized the chimp-hominid population. They used tools to some extent, but their primary adaptation was and is not tied to them.

Hominids, on the other hand, retained the ancestral body size and the ancestral mode of locomotion of the hominid-chimps. They coped with predation by running away into the savanna, climbing trees, and throwing things. But by the luck of the draw proto-hominids ended up in a much more arid environment than the proto-chimps. Selection was probably more severe on the hominids, and they evolved more rapidly, developing new ways to get at the abundant resources of the savanna. Tool use was one of the primary ways that hominids adapted. We investigate how and why tools would have been used in getting food in the next chapter, but their use strongly implies increased intelligence.

If tool making was indeed part of the initial suite of adaptations that the earliest hominids possessed when they set off on their evolutionary journey away from the apes, it should be possible with our current knowledge of brain function to postulate what some of the initial evolutionary changes in the brain would have been. The brain became larger, but which parts became larger and why?

The human brain is a complex structure, parts of which date back tens and sometimes hundreds of millions of years. The older parts of our brains function the same as do corresponding parts of crocodile, bird, rabbit, or monkey brains. What is different is the outermost part of our brains, the cortex, or covering, known as the cerebrum. Enlargement of parts of the cerebrum is the newest evolutionary development in the hominids. Anthropologists study this enlargement through the use of casts of the interior of early hominid skulls known as endocasts.

The earliest hominid partial brain endocasts known are those from Hadar and belong to *Australopithecus afarensis*. I studied one of these during a trip to the Ethiopia National Museum in 1987 during a review of *Australopithecus afarensis*. At issue was whether *Australopithecus afarensis* showed the advanced human position of a wrinkle in the cerebrum known as the lunate sulcus, as maintained by Ralph Holloway of Columbia University, or whether it showed the basic ape-like pattern of wrinkling, as maintained by Dean Falk at the State University of New York at Albany. The point is quite important in one sense. If australopithecines had the human pattern it would indicate that the front parts of the cerebrum, the areas that contain important functions like speech, had become enlarged and reconfigured as in modern humans. If not, australopithecines had much more ape-like brains. Most anthropologists and paleoneurologists have connected the ability to speak with the ability to make complex tools.

From my study of the partial skull from Hadar I concluded that the detailed measurements of the specimen that Holloway had taken were correct but that Falk likely had been right when she had criticized the orientation of the endocast, which had tended to give it a more human appearance. From my observations I would tend to agree with Falk that the australopithecines had an ape-like positioning of the lunate sulcus. From this admittedly rather limited evidence it is more likely that australopithecines lacked the ability to make complex tools and to speak. Falk does not see convincing evidence of the brain reorganization in the cerebrum until the appearance of *Homo habilis*, specifically Richard Leakey's famous 1470 skull, at about 1.9 million years ago. If the neurological argument is correct that the human pattern of lunate sulcus development presages both speech and complex tool making in hominids, then we might expect to see the appearance of stone tools near the time of the appearance of *Homo habilis*. This ability to use tools is what Louis Leakey had in mind when he and colleagues named "Handy Man" in the first place.

## The Beginnings of Stone Tool Use

When I began working in the Semliki valley of eastern Zaire, the oldest dated stone tool sites in the world were at Omo, excavated by Jean Chavaillon and Harry Merrick. They were situated geologically in members E and F of the Omo Shungura formation, dated to 2.2 to 2.4 million years ago. These crude quartz flakes and battered chunks were a full 500,000 years earlier than the Olduvai Gorge Bed I "living floor" that *Homo habilis* and his cousin *Australopithecus boisei* had occupied. But the Omo artifacts were small pieces unassociated directly with any fossil remains of animals or hominids. Although they were well-dated, they had no cultural context. In other words, it was not possible for archaeologists to tell what hominids did with the tools.

This mystery is what interested me the most. Why did early hominids, who had already lived in the savanna for 2 or 3 million years, suddenly begin to make stone tools? Gracile australopithecines such as *Australopithecus afarensis* and *africanus* apparently had been a successful group and apparently had done quite well without stone tools. If our deductions about the reorganization of the brain were correct, however, *Homo habilis* had been the first to systematically use rocks to fashion tools.

Other animals besides apes are known to use tools of some sort. That is, they employ objects outside their body for performing functions and manipulating their environments. The woodpecker finch on the Galapagos Islands uses a thorn that it plucks from a tree to pry out insects from under loose bark. The California sea otter cracks open mollusks using two smooth stones that it picks up from the sea shore. These species, like chimps, use tools for getting food. It is reasonable, then, to expect that early hominids must have begun to use tools for the same purpose.

Out on the African savanna there is a tremendous amount of meat on the hoof. In my years of surveying for fossils in savanna environments not unlike those of our early hominid ancestors, I saw many cases in which potential food was available for the taking: a hippo lying near-comatose in a pool by the river after having been mauled by a lion, a young wildebeest stuck in the mud and too exhausted to move, a clutch of eggs left by a secretary bird, a pair of fish eagles mating on the ground, a gazelle freshly killed by a lion, and grubs discovered in the soil during an excavation. All of these sources of protein could have been obtained by australopithecines, but only some of them could have been eaten without tools. Eggs can be easily broken open and licked off the hands, and grubs simply can be popped into the mouth. The gazelle kill would have been accessible to hominids because its carcass had already been butchered by the lions, assuming the lions, hyenas, jackals, and vultures had abandoned the carcass or allowed the hominids to approach it. But the other sources of meat would have been containerized and thus impossible for the hominids to get with their natural complement of teeth. We cannot bite through the hide of a hippo or an antelope, and even the skin of a bird would cause us problems. Wood, bone, or tooth implements, if they were used, would help a little, but they were not sharp enough. The teeth of other animals used as tools also are not particularly effective. Teeth cut by a scissor-like action in the jaws, slicing against the teeth in the opposing jaw. Hominids using a single jaw bone as a tool is a little like one hand clapping. It just doesn't work very well. If hominids had a way to cut open dead or dying animals that they found before the lions, hyenas, and other savanna meat eaters arrived on the scene, then they would have a significant advantage. This, then, is likely one of the major selective reasons for the origins of stone tool use. Stone tools became essentially an out-of-body set of teeth for early hominids.

Jean de Heinzelin found three stone flakes in his geological test trench in 1957 at a site called Kanyatsi on the northern shore of Lake Rutanzige in eastern Zaire. He was tempted to accept them as some of the earliest stone tools known, because they were in deposits that were Pliocene in age, at least 1.8 million years old. But caution prevailed, because extending the trench had revealed no more tools. Maybe they had fallen from higher in the sequence.

During the 1983–1984 field season in Zaire, de Heinzelin and I went back to Kanyatsi. Twenty-six years of erosion had slumped Jean's excavation walls, but new deposits had been exposed as well. We found a number of fossils and some stone flakes. Jean at first dismissed the artifacts, saying that they could have come from somewhere higher up in the sequence. But I collected them, and back in the Landrover we looked at them with a hand lens. Each of the artifacts bore the indelible reddish sandy sediment that is only found in the Pliocene-aged Lusso Bed ironstones. I thought this was excellent evidence on which to begin a new excavation at Kanyatsi.

When archaeologist Jack Harris arrived at the field site in Zaire, however, he was less impressed with Kanyatsi than with locality Senga 5A, and he decided to open an excavation there. The excavation team immediately began finding many stone tools and fossils in a good state of preservation. De Heinzelin's discovery of stone artifacts in the Pliocene of the Western Rift had not been a fluke. They really were there and, unlike Omo, they were directly associated with abundant animal fossils and fossil wood that would give us a good contextual understanding of the early hominids who had made the tools.

One of the things that impressed me the most was how similar the stone tools from Senga 5A were to ones that I had seen

Harry Merrick dig up at Locality 396 in the Omo in 1973. They were small and made from quartz. There were no large choppers or hand axes as found at Olduvai Bed I. Many of the Olduvai tools were made from other rock types, such as basalt, chert, or quartzite, stone that was more predictably flaked than large-veined quartz.

I had some experience with quartz. I remember as a boy reading about Louis Leakey's experiments in making stone tools and butchering goats to show how they worked, and I had tried it. I used the quartz locally available in stream beds where I grew up, and the experiment was a catastrophe. I could make nothing even vaguely resembling a chopper because the quartz exploded as soon as I hit it. I did end up with sharp little shards of quartz, and I cut up pieces of steak with these as an experiment. When I became an undergraduate, I was a research assistant in archaeology and became familiar with what the Indians did with quartz. I was mystified at the skill that it must have taken to fashion those delicate little arrow and spear points out of such an awful raw material. I concluded that the efforts of the early hominids at Senga had been much more akin to my first crude attempts at stone tool making with quartz than to the master American Indian artisans who had mastered the medium. Why might both the Omo and Senga tool makers have chosen quartz when other rock types were around and could have been used?

It occurred to me that the earliest, pre-Olduvai stone tool makers had been able to conceive of the idea of an out-of-mouth cutting mechanism, sort of like teeth that you held in your hand, but their technical ability to produce cutting edges had been simple. Really simple. They may have known only one way to get sharp-edged stones: throw rocks against other rocks on the ground and pick up the shattered pieces. Quartz, which explodes on impact, would have been perfect for this most primitive method of producing stone tools, whereas other rock types available were too hard to break. These early hominids had yet to discover the technique of holding one rock stationary and hitting it with another rock to pinpoint the force and break unveined crystalline rocks like basalt or chert.

The cranial capacity of adult *Homo habilis* was only around some 800 to 1,000 cubic centimeters, even less than that of my one-and-half-year-old son. It is difficult to conceive of how a toddler whom you see every day thinks, let alone how an adult hominid removed from you by 2 million years of time may have thought. But there may be some similarity. What my son lacks in adult experience he may make up in superior neurological wiring. What *Homo habilis* groups lacked in brain power they may have compensated for by repeated trial and error. I can imagine a toddler choosing the correct rock type and effectively throwing it on the ground with sufficient force to produce sharp-edged fragments. It is much harder to imagine him holding a rock between his feet or in one hand and delivering a controlled blow with a second rock. This ability will develop later and may also have come along in the later evolution of *Homo habilis*.

Once *Homo habilis* had sharp stones, what did they do with them? At Senga 5A Jack Harris's team found one intriguing answer. The excavation turned up a large tortoise shell that had incised lines on it. A number of things can produce incisions like this in bones in fossil sites. Carnivore teeth raking across bones as they tear off muscle can leave long depressions. Hooves of herbivores trampling over a bone can scratch the surface of the bone. And sometimes the sharp edges of stone tools that cut too deep can leave cut marks on bone. Analysis of the microscopic characteristics of the incisions and a consideration of where they occur can help to answer the question.

The Senga 5A tortoise shell had sharply incised, regular channels that looked very much like stone tool cut marks. Tooth marks from carnivores are more rounded in cross-section. Trampling marks tend to be rounded in cross-section and irregular along their length. They also tend to be clustered with many other scratches. The Senga cut marks, by contrast, were isolated. Finally, our fossil tortoise expert, Peter Meylan, weighed in with his observations on tortoise anatomy. The location of the cut marks was where ligaments hold the skin of the back leg of the tortoise to the shell. He had seen modern turtle butchers make cuts exactly in this place to open up the shell.

We accepted that the Senga early hominids used their stone flakes to cut open animals for meat. A tortoise made perfect sense. It was slow and so could easily have been caught by early hominids. It was impervious to attack by lion, hyena, leopard, cheetah, jackal, and vulture because of its shell and thus would have been a meal that promised less interspecies competition. What we had found appeared to be the earliest evidence for the use of stone tools for meat eating.

## TOOL MAKING, CULTURE, AND THE BRAIN

The advent of stone tool making is a momentous development in the hominid career. It marks the beginning of culture as the primary adaptive mechanism for hominids, even if it was what I have called *bas culture,* an earlier version of the software. But just as an understanding of the software programs that run on a particular computer can give clues as to the configuration of the hardware, so too can an understanding of early culture provide clues as to the brain that produced that culture.

Nick Toth of Indiana University has studied early hominid stone tool making from a technical standpoint. By assembling the flakes back into the original core from which they were made, he has determined that the vast majority of early hominid tool makers were right-handed. This discovery is significant because it tells us a lot about the brain. Only humans have a lateralized brain—that is, one in which the cerebral hemispheres have different jobs and different aptitudes. The left side of the cerebrum is generally where language ability is located, along with logical planning, numerical ability, and tool-making ability. The right side is where abstract thought, musical ability, and three-dimensional problem solving are found. The left cerebral hemisphere controls the right hand and whole right side of the body, and the right hemisphere controls the left side. If early hominids showed the same statistical preponderance of right-handedness over left-handedness as modern people, then we can surmise that they had the same sort of lateralized brain function that we do.

Understanding how the brain functioned in early hominids is a critical component to understanding how our early ancestors behaved and thus critically important to understanding how we behave. But fleshing out a model for the mind of the missing link is tricky business. We have only a few endocasts, stone tools, and comparative studies on modern brains to draw upon. Yet we must make the attempt to put the whole into some sort of testable framework to organize our observations and to direct future research.

Based on an extrapolation backward in the trend of hominid brain growth to body size over time, I postulate that the enlarged growth of the hominid brain is an attribute that marked the beginning of the hominid branch from the apes. Another anthropologist, Robert Martin of the Anthropological Institute in Zurich, came to nearly the same conclusion from a consideration of comparative growth rates of the brain in non-human primates and other mammals. Martin found that, unlike other mammals, human infants continue a fetal growth rate of the brain even after birth. It is as if the brain needs to grow so much that the maximum gestation period is not enough. Birth occurs because the mother's pelvis cannot accommodate passage of such a big-headed infant if brain growth were to continue in utero. So human infants are born in a state of almost total helplessness, known in biological parlance as "altricial." Early hominids would have experienced a similar period of infant dependency because they also had increased brain size, although not to the extent seen in modern humans. Martin believes that this altriciality and increased brain size may have characterized the earliest hominids.

I have suggested that the driving force behind the selection for large brain size is the use of tools as a food-getting mechanism. There is little direct evidence for reliance on tools by the earliest hominids currently known, the australopithecines, although use of bone has recently been postulated in the South African early hominids. But the appearance of stone tools at about 2.5 million years ago corresponds to a human pattern of brain organization in the temporal lobe of the brain, as evidenced by the *Homo habilis* endocasts. By analogy with modern human brain structure, this finding may mean that early *Homo* could talk in a rudimentary way. Certainly they could not have produced Linnaean Society types of debate, but verbal communication in early *Homo* was likely quite different from the verbal communication of apes. Anthropologist Grover Krantz made the interesting suggestion that early hominids may have learned language at a much later age—say, age twelve. They probably had a much slower learning curve than modern humans. Talking may have approached the time of puberty in the earliest hominids; there might have been cases in the earliest hominids in which individuals procreated before they could talk!

There are many hypotheses on why the hominid brain grew so rapidly. Some have suggested competition in outwitting carnivores in hunting, or adaptation to social cooperation with greater group cohesion, or the acquisition of language, or, as here, the use of tools in food getting. Dean Falk recently suggested that the reorganization of cranial veins in a bipedal hominid allowed more effective cooling of the brain and thus permitted the evolution of increased brain size. Many of these theories are not mutually incompatible. The evolutionary forces that they invoke may have been real, but they may not have affected hominid evolution at the times that their proposers suggest, or they might not have acted in quite the degree that they suggest. Hominids, for example, may have competed with carnivores by using tools. Or, for example, early hominid females may have cooperated in a more cohesive fashion by using tools for gathering food.

We will only be able to decide between the many theoretical possibilities for the evolution of the hominid brain if we have an accurate setting for our models, which then can be tested. I believe that the best approach to establishing a context for the evolution of the adaptations specific to hominids is to look generally at their environment. This pursuit has led me into ecology, and specifically paleoecology. Paleoecology is the enterprise that attempts to reconstruct lost worlds, putting together all the clues about a species' habitat, diet, population statistics, demographics, reproduction, and behavior. This environment is where our pursuit of the missing link takes us next for its next-to-final stop.

# 9

# Climatic Change and the Hominid Diaspora

When Raymond Dart first described *Australopithecus africanus* in 1925 from the South African site of Taung, he emphasized the uniqueness of the environment in which this strange, small-brained little hominid had lived. It was in what South Africans called *veldt* in Afrikaans, a cognate of the word field in English. The environment could have also been termed prairie, a word from French meaning meadow that Anglo-Americans used to describe the extensive plains in the American Midwest, or *pampa*, a Quechua Indian word used in Spanish for the grasslands of Argentina and neighboring areas of South America. But the term that stuck was savanna, a word from the Caribbean Indian language of Taino that was used by the early Spanish colonialists in Florida to refer to plains in the southeastern United States. Savanna was appropriated by ecologists and applied to Africa. The term refers to plains with a few bushes and spindly trees grading into more wooded but still open habitats. On both sides of savanna are steppe, referring to totally treeless semidesert, and woodland, referring to terrain in which the clumps of trees are more abundant than the spaces of grass between them.

## THE SAVANNA MYTH AND HOMINID ORIGINS

But whatever name was applied to its environment, Dart thought that *Australopithecus* had lived in a distinctly different habitat than the living African apes. Taung was near the Kalahari Desert and is quite dry today. Dart and others after him concluded from the other animal fossils collected from the site that it had had a similar climate in the past when *Australopithecus* lived there. Dart contrasted this largely tree-less grassland with the dense forest habitats of the gorilla and chimpanzee, and he used this argument to bolster his attribution of the Taung skull to a new genus and species of African hominoid. Thus was born the savanna hypothesis of early hominid ecology.

Dart's hypotheses about *Australopithecus africanus* met strong opposition in scientific circles, particularly regarding the anatomical characteristics that according to him showed that Taung was closely related to the human lineage. But except for a desultory comment from Sir Arthur Keith that Taung may well have been a "jungle," Dart's claim concerning the savanna habitat of Taung went essentially unchallenged. Perhaps this was because the idea that humanity originated in a grassland habitat had been proposed some years earlier by none other than the codiscoverer of the theory of evolution by natural selection, Alfred Russel Wallace.

Wallace worked largely in Australasia and not surprisingly had a somewhat Asia-centric viewpoint. He started with the proposition that hominid bipedalism was clearly an adaptation for walking in flat areas. The most extensive areas of flat terrain in Eurasia were the steppes of central Asia, and Wallace guessed that this region may have been the site of hominid origins. Although Dart had changed the geographical component of Wallace's thesis with his discovery of *Australopithecus* in, of all places, South Africa, the environmental aspect of the argument was familiar territory.

I am always skeptical of ideas in science that have been

accepted uncritically for a long time. They begin to take on mythic proportions. So in graduate school I began to investigate what I conceived of as the savanna myth of hominid origins. I embarked on this study not so much to debunk Dart's hypothesis as to discover what the habitats of early hominids actually were. By the early 1970s this question had become more than a little confusing.

## ECOLOGICAL EXCLUSION AND THE SINGLE SPECIES HYPOTHESIS

By the end of the 1960s the discoveries documenting two species of hominids living side-by-side at Olduvai 1.8 million years ago had created a theoretical crisis in anthropology. Here were two hominids living in the same environment, at the same time, in an apparently cozy ecological arrangement that lasted for a million years. Clearly Darwin's hypothesis about tool use and reduction of the canine teeth had to be modified. Both *Homo habilis* and *Australopithecus boisei* had small canines, but the paradigm that most anthropologists believed did not allow them to accept that both groups could both be tool using and therefore culture bearing. Anthropologists had generally believed since Ernst Mayr's paper at the 1950 Cold Spring Harbor Symposium that only one culture-bearing hominid species could exist at any one time because of an ecological theory known as the competitive exclusion principle. Species coexisting in an area had to eat different foods, sleep in different places, and generally make a living in a different way than other species. Otherwise, two similar groups would use up the same environmental resources and eventually one of them would become extinct. Culture was such an adaptable mechanism that it was impossible to think that two hominids could have used it and been able to coexist. In ecological terms, two hominid species could not coexist in the same cultural ecological niche. What, then, were anthropologists to make of two hominid species coexisting at Olduvai?

For a number of years some anthropologists simply denied the existence of two species of early hominids. Loring Brace and Milford Wolpoff at the University of Michigan developed what was known as the single species hypothesis. They maintained that what other anthropologists considered two species were in fact males and females of the same species. Not until complete early *Homo* and robust australopithecine skulls, which allowed clear species diagnoses, were described in the late 1970s at east Lake Turkana did Wolpoff and Brace give up the single species hypothesis. But a theoretical void was left because still no explanation was generally accepted as to how two hominid species could coexist.

Whereas Wolpoff and Brace had used the anatomy of the fossils to argue against the presence of two species in the early hominid fossil record, other anthropologists who accepted the presence of two species had proposed hypotheses based on ecological principles to explain the coexistence. One explanation by Todd and Blumenberg in 1974, for example, held that there was a mutually beneficial relationship between early *Homo* and the robust australopithecines, ecologically termed a symbiosis. Early *Homo*, a small-bodied tool user, had hunted meat and brought it back to camp. Robusts had been large-bodied non-tool-using hunters, foraging for food around the periphery of the early *Homo* group. Both species hypothetically benefitted from this arrangement. Scraps from early *Homo* meals could be scavenged by the robusts, who could chew tougher food items because of their massive molar teeth and jaws. For the early *Homo* group, the presence of the burly robusts helped to ward off predators.

This idea was clever, and it was plausible, but it was getting close to the just-so story category. I wanted to devise tests of at least some of these ideas and bring empirical results to bear on the problems. I focused on the ecological data available from the East and South African early hominid sites and then I went into depth on some of the tests using data from the Omo.

## DIET AND HOMINID EVOLUTION

Starting from first principles, I reasoned that the most important aspect of early hominid ecology to know is diet. Species are defined in large part by what they eat, and the feeding adaptations of early hominid species had to have differed, judging from the major differences in their teeth. Early *Homo* had large incisors and small cheek teeth (the premolars and molars), whereas robust australopithecines had diminutive incisors and huge cheek teeth. Both species had small canines. Zoologist John Robinson proposed the dietary hypothesis, in which he suggested that the robusts were vegetarians and the early *Homo* and their gracile australopithecine ancestors had been omnivores, eating meat and everything else they could get their hands on.

I used Robinson's hypothesis as my jumping off point. If we assumed that the robusts had been total plant eaters and that early *Homo* had been omnivorous, there should be predictable indications as to how their diets had differed.

The first way that I predicted early *Homo* and robust australopithecines would differ was in the strontium contents of their teeth. Two paleontologists, Michael Voorhies and Heinrich Toots, had reported that fossils of herbivores, like deer, had much higher values of strontium than carnivores did. This variance is because strontium becomes diluted as it moves up the food chain. The strontium in ground water is diminished in the plants that take it up; the herbivores who eat the plants have lower strontium values; the primary carnivores who eat the herbivores have even lower levels of strontium; and large carnivores have the lowest strontium levels. If Robinson's dietary hypothesis was valid and the strontium system worked, early *Homo* should have substantially lower strontium values than robust australopithecines.

The first problem that I had to overcome was destruction of the fossils. Voorhies and Toots had ground up their fossils and done very accurate measurements. I certainly was not going to be able to grind up any fossil teeth, especially those of early hominids. So with a master technician in the Berkeley Geology Department I developed a nondestructive technique using a system of analysis known as X-ray fluorescence. In this technique you put a specimen in the chamber of an X-ray fluorescence machine, zap it with an X-ray, and read out the counts of each chemical element in the specimen. The elements in the specimen give off energy in wavelengths specific to those elements when they are hit with the X-ray. In this way you can tell what elements are in the specimen. I looked at strontium. We developed a little metal shield that we taped onto each specimen to ensure that a standard surface area of each tooth specimen was read by the machine.

As soon as we had worked out the bugs in the system, I did a pilot study. The method seemed to work. A fossil hyena tooth had the lowest strontium. Hominids were next lowest. A tooth from an antelope, which grazed on grass, was high in strontium. And a giraffe tooth was the highest in strontium. Earlier researchers had predicted that animals that browse on leaves, as giraffes do, would be very high in strontium. So we started off on the project with great expectations. I analyzed a total of 2,865 teeth, including 68 hominids. After many hundreds of hours over several months in the X-ray fluorescence lab I tabulated the results. I remember the confusion and disappointment that set in as I graphed the results. The carnivores did not appear the lowest in strontium. Strangely, they were higher than the herbivores. And hominids appeared high too, but all bets were off regarding diet reconstruction. If the method did not work with animals of known diet it could hardly be trusted to indicate anything about unknown diets.

When the initial shock wore off I realized that the levels of strontium that we were finding in the teeth were the same as in the enclosing sediments from which the bones were excavated. They were much higher than would be expected in animal tissues. Apparently ground water, slowly percolating through the teeth over millions of years, had not only fossilized and mineralized the teeth but had replaced the in-life values of strontium. We published a paper presenting our disappointing results, and I moved on to another method of reconstructing diet. However, this story has a sequel, provided by paleoanthropologist Andrew Sillen of the University of Cape Town. Sillen persisted with studying strontium in fossil animals and found that the variability in strontium values is greater in living animals than first believed, based on the specifics of what they eat, and this variability can confound interpretation. He also found that contamination could be controlled by looking at nonbiological elements that had definitely come in from the surrounding rocks. Contaminated specimens could then be eliminated from the study. Sillen has not yet come up with definitive answers to early hominid diet, but the technique may yet yield the answer.

Meanwhile I turned to a more direct approach to investigating early hominid diet. Coprolites (from the Latin *copros*, feces, and *lithos*, stone) are fossilized dung. They are quite common in many fossil hominid sites, including Omo, Olduvai, and Hadar. As a graduate student I became interested in the tremendous amount of data on diet that could be obtained from coprolites if you could tell which animal species they came from. I spent quite a bit of time studying the Omo coprolites—measuring them, X-raying them, sectioning them, and taking many scanning electron (SEM) micrographs of objects inside. But I had a very hard time running down any information on coprolites from other sites. Paleontologists, it seems, have not spent an inordinate amount of time investigating the paleobiological significance of coprolites. I can understand why after suffering through the obligate bad jokes that resulted when I told my friends and colleagues what I was studying. It must be hard to resist mock enthusiasm ("No shit!"), or acted-out third-person commentary ("Boaz's work is really shit"), followed by uproarious laughter. One of my friends even penned some doggerel and gave it to me:

> Little birdies, little birdies
> I can tell them by their turdies.
> Some are red and green and blue.
> Little turdie what are you?

But I began to find very interesting things in the coprolites under the SEM, most of which I could only vaguely identify. With very little past work to go on, I was forced to rifle the library shelves for comparative SEM photos of everything from pollen grains to sand particles to meat fibers to tapeworms. One of my first identifications was a section of a chewed up leaf, magnified several hundred times. I found it in an elongated coprolite from a robust australopithecine site at Omo that from its shape could easily be from a hominid. I was ecstatic.

Here was a 2-million-year-old, totally rock-like coprolite that preserved in perfect form a structure as delicate as a leaf. Could this be the first real evidence that robust australopithecines were vegetarians? I had a lot of ground to cover before I could make any statements like that, but I felt that I was well on my way.

Anyone who has gone hiking or camping or has been around a barnyard can probably take the first steps toward coprolite identification. Dung of herbivorous animals such as deer, sheep, or cows is pelletized and very different from the dung of carnivores, such as dogs, cats, or raccoons. Human feces are very similar in form to carnivores. This is the first problem in identifying hominid coprolites. Herbivore coprolites can be easily sorted from carnivore-like coprolites on the basis of shape. But within the carnivore-like group, hominids or primates in general cannot be sorted out on this basis. Modern leopard dung can look just like human dung.

A colleague of mine, Frank Spencer, and I undertook a study of modern carnivore scats to determine how they might be able to be distinguished from humans'. We were able to obtain scats from carnivores at the Bronx Zoo and the National Zoo in Washington. We were unsure whether the diets of captive animals would be very useful in assessing wild species' diets, but in the case of the carnivores the telling difference from humans was not dietary per se, it was hair. Lions, cheetahs, leopards, and dogs groom themselves by licking their fur, which they then swallow. Hair is not digested. It goes through the digestive tract and is deposited in the feces. In coprolites the hairs are gone, but hundreds of small-diameter holes remain. In looking back at the coprolites, my supposed robust australopithecine coprolite from Omo had hundreds of little holes in it. It was a large cat, not a hominid.

We are now working on how to sort nonhuman primates from hominid coprolites, and there are two promising methods. The first is particle size within the coprolites. Modern humans have a very small particle size in their feces. Everything we eat is ground up into very small sizes either by processing before it is cooked, cooking itself, or by chewing. Baboon and gorilla dung collected in the wild, however, shows that these species do not chew up their food into such small bits, and this characteristic is reflected in the particle size in their dung. Pre-fire-using hominids would show less difference from baboons and apes, but the structure of their crushing molars indicates that they probably shared modern humans' approach to making food digestible: crushing it into very small bits.

The coprolite studies are still underway, but the initial results prompted me to question Robinson's dietary hypothesis for early hominids. There is no compelling reason to believe that the large-molared robust australopithecines were strictly herbivorous. Their thick-enamelled molars more likely were for crushing food into small digestible bits than for cutting or shearing plants. They may likely have eaten a lot of plants, but nothing would have precluded them from chewing up meat, which is also very fibrous and difficult to digest. The question then comes down to: why didn't the robusts have a competitive advantage over the puny-jawed early *Homos* and drive them to extinction? There are a number of potential answers, but without further research we will not know for sure. The most likely answer may be tools. Early *Homo* may have specialized in a nondental way to cut up dead animals that robusts could not catch, kill, and eat, or to dig up underground food resources that robusts could not get to, or to knock down certain fruits from trees that robusts could not climb, and so forth.

Another telling blow to the dietary hypothesis was in my mind the relative numbers of robust australopithecines and early *Homos* that turned up at fossil sites. Ecologists early on discovered that there is a pyramid of numbers when it comes to population and the food chain. For example, there are many, many mice in a given area of African savanna at the bottom of the pyramid, fewer snakes that prey on the mice, fewer mongooses who prey on the snakes, and, at the top of the pyramid, fewer leopards who prey on the mongooses. Herbivores, even large-bodied herbivores like antelopes, are much more common than the carnivores that prey on them. If robust australopithecines had been herbivores, the numbers of fossils from sites where both they and early *Homo* were found should significantly outnumber *Homo*. I compared all the hominid numbers from the known sites and there were about equal numbers of robusts and early *Homo*. Herbivorous baboons, however, are quite a bit more common, in Omo and other fossil sites in this time period. By looking at numbers of fossils, then, robust australopithecines appeared very similar to early *Homo*. Their rarity implies that they were more similar to carnivores in dietary habits than not, but other reasons might be behind their small numbers.

## HOMINID POPULATION NUMBERS AND ECOLOGY

The rarity of hominids in the fossil record is legendary. I have spent twenty years looking and I fortunately have at least a few scraps of bone to show for it. Hominids are as rare as other very rare components of fossil faunas in the Pliocene-to-Pleistocene time periods, such as carnivores, aardvarks, camels, and birds.

I wanted to know why hominids were rare. This question necessitated finding out why all the rare animals were rare.

Trying to understand the relationships of population numbers to diet in the past is a question of ecology. We would like to know what an average day in the life of the Pliocene looked like by reconstructing an assemblage of fossil animals. But we cannot assume that an excavation of fossil bones necessarily preserves a snapshot of former life, comparable to Pompeii suddenly being covered by the ashes of Mount Vesuvius. Sometimes this is clearly the case, but at other times bones of animals may be far removed in time and space from where their original owners lived and died. The field of paleontology that deals with this interconnecting discipline between ecology and paleontology is called *taphonomy*.

The difference between looking at the question of the rarity of animals in the fossil record from a purely ecological perspective versus one tempered by taphonomy is that in the first instance you assume that rarity as a fossil means past rarity in the environment. This is not necessarily the case. Take birds for example. On an average morning when *Homo habilis* woke up, the first sounds and sights would have been birds, yet they are among the rarest of fossils. This is because bird bones are fragile. They are hollow to allow flight. Thus if a bird carcass is chewed up by a carnivore, usually nothing is left to fossilize, and even if there are some remains the weathering effects of sun and rain finish them off. Small-bodied animals are the most susceptible to this sort of differential destruction. Small animals are called *microfauna*. To find them we wash tons of sediment through fine screens and then pick through and sort the dried sediment under microscopes. To find just a few teeth of the small rodents, insectivores, bats, and small primates takes hundreds of hours of sorting. But these species were major parts of the past ecology.

Another major reason that animals may not turn up in a fossil site even though they were around in some numbers is that they didn't live exactly where the fossils were deposited. At Omo, for example, geological evidence shows that all the fossils were buried by the action of the waters of the Omo River or the lake into which the river flowed. If animals such as camels lived far away from water, they would have very little probability of making it into the fossil bone assemblage when they died. The example of camels is obvious, but this same effect may have been at work more subtly in affecting the rarity of other animals. In my analyses I found that colobus monkey fossils are quite rare at Omo, about as rare as hominids. Although they are not large animals, some of them reach the size of baboons, which are quite common in the Omo deposits. If modern African forests are any guide, colobus monkeys were common in the environment, and they should have lived in the trees next to the river. When they died, their bodies should have fallen into the river and been deposited. I was quite confused by the rarity of colobus monkeys until I read that they never descend to drink water; instead they derive all their needs from the leaves that make up their diet. If colobus monkeys never touch the water, they are in a sense as distant from the river as the camels. I thus explained the rarity of colobus monkeys on the basis that although they lived in the forest fringing the river, they were not living exactly on the river edge where they would have made it into the fossil deposit.

To understand how hominid fossils fit into this picture of a fossil deposit laid down by the ancient Omo, I wanted to understand how human bones move in flowing water. At Berkeley I undertook human taphonomy experiments with Kay Behrensmeyer (Chapter 3). We used an artificial stream channel, known as a flume, in the Engineering School. The flume had been built by Albert Einstein's son, who had used it to study flow dynamics in streams. We found out which bones were easily moved away under varying current speeds and which stayed on the bottom. With the flume set at the probable speed of the ancient Omo River we defined a group of bones that would be swept away, a "transport" group, and a group that stayed on the bottom, which we called the "lag" group of bones. The lag group was composed of exactly the body parts that we found at Omo: single teeth, finger and toe bones, jaws, vertebrae, some limb bones, and broken skull fragments. A surprising discovery involved how complete skulls behaved in the flume. No matter how water-logged a human skull became and no matter in which orientation it was dropped into the water, it floated. This was why there were no skulls in the Omo hominid fossil collection except the L894 *Homo habilis* skull that Howell and I described from the lake beds of Member G. The engineers who watched us conduct the study were very interested in the floating skull. They called the campus newspaper, and a photograph of the phenomenon of our floating human skull in Einstein's son's flume was soon famous at Berkeley. The flume study confirmed that the Omo fossils in large part had been winnowed by fast-flowing river water before they had been covered up and buried as fossils.

The flume study answered another question that was very important in assessing the causes of rarity of bones in fossil sites. Size of the bone alone did not predict whether it was swept away by the water. The important factor was density of the bone. Large, light bones like the shoulder blade might be easily swept away, but small dense elements like teeth were virtually immobile. This difference also meant that bones of large animals would not necessarily be disproportionately represented over those of other animals, except for the very small microfaunal species. All the dense bones of animals would be deposited regardless of size of the animal. This discovery was quite exciting to me because it meant that the numbers of fossil bones that I had excavated at Omo might represent the real relative numbers of the species in the environment.

Hominid fossils, all single teeth, make up slightly less than one percent of all the identified mammal fossils from my excavation at Omo Locality 398. The implication from the flume study is that hominids were truly rare in the environment. Their bones would not have been disproportionately underrepresented simply because they were on the smaller end of the size spectrum. The water requirements of early hominids, I feel sure, were comparable to humans and chimps, who drink water each day, and thus they would have been closely tied to the river and lake at least at some point during their daily rounds. But rare is a relative term, I wanted to know how rare.

I was able to estimate the population densities of early hominids by using the data from the Omo excavations and comparing population densities of modern African mammals to the values their fossil relatives had in the excavations. If the fossil hominid-to-pig ratio, for example, was one-to-ten, I multiplied the modern population density of pigs in African game parks by that number and derived an estimate of early hominid density. I did the same for hominids to hippos, hominids to antelopes, hominids to elephants, and so forth. The fossil ratios of other animal groups, say antelopes to elephants, compared well with the modern ratios of game park population densities, so the indications from the flume study checked out and gave me confidence that the hominid population estimates were close to real values. Hominids came out to between 1.0 and 0.1 individual per square kilometer, about the same density as modern human hunter-gatherers like the South African San or Bushman.

This finding has profound implications. It means that early hominids ranged over a very large territory, relatively larger for them than for modern human groups because they were significantly smaller in body size. It confirms that bipedalism had to function as an effective long-distance transport mechanism to cover such a large territory on the open savanna. And it tied into diet by implying that whatever the hominids were eating, it was widely dispersed. Ecologically speaking, animals with high population densities and small ranges, like antelopes, eat easily obtainable but low-quality food like grass. Carnivores, on the other hand, have large range sizes and eat high-quality food that is much more difficult to obtain, like antelopes. Hominids clearly fell ecologically into the latter group.

## ECOLOGICAL CHANGE AND HOMINID EVOLUTION

Because the Omo data were so well controlled over such a long time period, they are of much theoretical interest to anthropologists interested in how ecological change affected hominid evolution. The Omo study of fossil pollen, undertaken by Raymonde Bonnefille of the University of Marseille, was widely cited because it showed a change to widespread grasslands at about 2.3 million years ago. Hank Wesselman's doctoral dissertation on the Omo microfauna showed there were more open-country rodents around about the same time. There were more woods and fringing forests early in the Omo record than later, after about 2.5 million years ago. It was widely pointed out that this environmental change corresponded closely to the date of appearance of *Homo habilis* and the robust australopithecines in eastern and southern Africa. Could there be a connection?

There could have been several connections. You may remember that I argued in Chapter 6 that the onslaught of the savannas created conditions for the two hominid-ape splits: hominids could walk across the plains, leaving first the gorillas, then chimps, behind. What about the difference between *Homo habilis* and other hominids? Let's first look at those strange parahumans, the robust australopithecines. In my model of the ecology of the early hominids, I postulate that the robust australopithecines had to have been geographically cut off from the evolving *Homo* lineage during its formative years. We have in eastern and southern Africa good ancestors for *Homo*, but no good ancestors for the robusts. I suggested that they evolved somewhere else first, isolated from

*Australopithecus afarensis* and *africanus*, the ancestors of *Homo* in East and South Africa. But where?

The robusts thrived and apparently were well adapted to savanna environments when they did get into East and South Africa. They appear abruptly in the Omo and west Lake Turkana at 2.5 million years ago, corresponding to the spread of grasslands. The only other large region that was savanna in the early Pliocene when the robusts differentiated was northern Africa. I think that this region was likely where the robusts evolved from a woodland or forest-fringe hominid not unlike *Australopithecus afarensis*, but they remained separated from their cousins by the line of dense woodlands and forests that stretched from the Central Forest Refuge of Zaire across southern Ethiopia and into northern Kenya. Zoologist Jonathan Kingdon found that the forest animals that live today in isolated forests on the coast of Kenya arrived there from central Africa by dispersal through these interconnecting forests. When this forest connection was broken up by intervening savannas 2.5 million years ago, the robusts were able to extend their range into eastern and southern Africa. Vast areas of the Sahara await future paleoanthropologists to test this hypothesis. We would expect the sites earlier than 2.5 million years in northern Africa to have only robust australopithecines. Then early *Homo*, in a dispersal event converse to that of the robusts, would have been able to extend its range northwards.

I am struck with two overarching facts shown by the deep sea core record of climate change. First, average global temperature has been falling for the past 5 million years or so. The large-scale pattern of the curve shows that cold temperatures and aridity overall have increased. Second, the small-scale pattern of change has shown progressively wider and wider fluctuations. The cold periods get colder and drier and the warm periods get warmer and wetter. These trends have had a tremendous effect on hominid evolution and on the dispersal of the hominids out of Africa to populate the rest of the world.

In several papers I and my Lamont-Doherty coauthors proposed the mechanism of a "climatic pump" that worked periodically over the last several hundreds of thousands of years to eject hominids from their African homeland. The 2.5-million-year event primed the pump, allowing early *Homo* to spread into northern Africa from eastern and southern Africa. As the environment became progressively drier, food and water became scarcer and the Sahara savanna became the Sahara Desert; hominid populations north of the advancing arid conditions were pushed farther and farther north toward the Mediterranean. Some populations escaped North Africa and expanded into the more temperate areas across the Isthmus of Suez—Eurasia.

Some paleoanthropologists, particularly in France, believe that even at the early 2.5-million-year date early *Homo* may have been pushed into Europe. Some early stone tool occurrences in the Massif Central of France are dated to greater than 2 million years ago and may record a hominid presence. But without hominid fossils to confirm such a momentous expansion, most paleoanthropologists remain skeptical.

During periods of returning warm, wet conditions, hominid populations re-expanded in northern Africa, with renewed influxes of populations from sub-Saharan Africa and possibly even some migration back from Eurasia, as some paleoanthropologists who work in Asia think. The hominids stayed there until the next, more severe bout of climatic aridity struck, on the average of about every ten to fifteen thousand years. Hominid populations in northern Africa would then have been pumped into Eurasia, where conditions were cold but where they could at least find water and large animals for food.

## PALEOCLIMATE AND GENETIC EVOLUTION

Each climatic cycle and its attendant migrations of hominid populations accounted for the mixing of gene pools. This gene mixing accounts for the observation that evolutionary advances seen in the fossil skulls of hominids tended to take place more or less simultaneously over the entire Old World. There is an evolutionary unity to the human species across the broad front of its evolutionary advance. But there was a high cost for this unity: widespread extinction of human populations caught in unfavorable conditions and unable to adapt or migrate to better areas.

Arid phases of climatic change during the Pleistocene Epoch undoubtedly stressed hominid populations. With little water and no food, there must have been fierce intergroup competition at times. There was probably widespread local extinction, in both northern Africa and in Eurasia, where conditions were harshest. Sub-Saharan Africa remained the font of hominid repopulation. Today sub-Saharan Africa retains four-fifths of the genetic diversity of the human species, a point that I will return to below.

One indication of the scale of the extinctions and the bottlenecks through which human evolution passed during these times come from genetics. The structure of human chromosomes shows that there are a number of places where large pieces of the chromosomes were detached and repositioned since the time of the human split from the chimps. An animal that has undergone a chromosomal change of this magnitude cannot produce viable offspring with a mate that has the normal chromosome structure. The chromosomes from the two parents will not match correctly and the mother will have a spontaneous abortion. For one of these chromosomal rearrangements to become fixed in the population, closely related individuals who share the trait must mate. Population genetics tell us that, mathematically, the effective population would have to be about twenty adult individuals when this sort of genetic reorganization happens. We don't know where along the lineages of humans and chimps these chromosomal rearrangements occurred, but to be conservative let's say that four happened on the way to chimps and four happened on the way to humans. This supposition would mean that four times in human evolution the human species was twenty people away from extinction.

Population bottlenecks likely correspond to the periods of climatic aridity that we see in the paleoclimatic record. There are two well-known population bottlenecks in human evolution and both are explicable in terms of paleoclimatic changes.

The first bottleneck concerns what has been termed the

African Eve hypothesis. Molecular anthropologists Allan Wilson and Becky Cann of the University of California at Berkeley studied the DNA in small structures in human cells known as mitochondria. The DNA in mitochondria is interesting because it evolves about ten times more rapidly than the main DNA in the cell nucleus. Wilson and Cann thought that mitochondrial DNA, abbreviated as mtDNA, would be a good molecule to look at because it could reveal a great deal about more recent patterns of human evolution. Their research created a furor in the world of paleoanthropology.

Wilson and Cann found that most mtDNA diversity is found in Africa—not a particularly surprising discovery to anthropologists with even nodding familiarity with human genetic diversity. They thus rooted in Africa the evolutionary tree that they had constructed from the mtDNA data. When their results were in, they found that all modern human mtDNA can be traced back to an origin about 120,000 to 200,000 years ago. Taken to its ultimate source, they suggested that all humans alive in the world today evolved from one mother who lived in Africa between the dates of 120,000 and 200,000 years ago.

Wilson and Cann's findings are very important, but their conclusions mistake lineages of genes with lineages of organisms. Their conclusions should accurately be stated as "the molecular configuration of the mitochondrial DNA of all living humans derives from the mitochondrial DNA of one mother who lived in Africa between 120,000 and 200,000 years ago." This statement is quite different from saying that all humans evolved from one mother. For those of us who have worked closely with molecular biologists and have a fondness for them, the misstatement by Wilson and Cann is understandable. Because they focus on the molecular level, they do not always recognize the organismal level. From their point of view, molecules can easily become the same as the whole animal. Many anthropologists have accepted this viewpoint and moved on. Milford Wolpoff did not.

Wolpoff restated the mtDNA results to mean that all hominid fossils earlier than 200,000 years ago outside Africa were irrelevant to understanding human evolution. He and colleagues went to great lengths to show that there were traits in Asian, European, and Australian hominid fossils that showed continuity from early *Homo erectus*-stage hominids to modern *Homo sapiens*. For example, modern Asians have little ridges on the insides of their upper incisor teeth and so did Chinese *Homo erectus*. Wolpoff maintains that anatomical traits like these "shovel-shaped incisors" show that regional continuity rather than replacement by African populations was the pattern of hominid evolution.

Wolpoff, however, made a mistake of interpretation similar to that of Wilson and Cann. He confused lineages of traits with lineages of organisms. For Wilson and Cann, an organism is represented by a molecule. For Wolpoff, an organism is represented by a tooth wrinkle or a brow ridge configuration. Both may be equally right in their own frames of reference, but both are equally wrong when we want to understand the overall patterns of hominid evolution.

It is eminently possible that a human being living in downtown Shanghai has mitochondrial DNA from a distant ancestor in Africa as well as shovel-shaped incisors inherited from early Chinese *Homo erectus*. Human beings are mosaics of evolutionary innovations as well as evolutionary conservatism. We inherit our hand structure from very ancient early amphibians, yet the brain mechanisms that control the muscles of the hand are very recent in evolutionary time. It is therefore not surprising that different traits in one individual or population, be they genetic or morphological, can come from different sources. Those traits, looked at apart from the whole organism, can lead to faulty conclusions. The fossil morphology shows that there was some gene exchange between indigenous populations and those who migrated in.

What the mtDNA results show is that there were significant population movements out of Africa at the beginning of the period when we see for the first time fossils that represent archaic *Homo sapiens*. Paleoclimatic data indicate that a long period of cold temperatures, known as Oxygen Isotope Stage 6, began just after 200,000 years ago and extended to about 130,000 years ago. This period of increased aridity in

Africa would have been the mechanism that pushed African populations into Eurasia. It corresponds well with the postulated molecular date for the African origin of modern human mtDNA.

Another major area of debate in paleoanthropology that climatic data has helped to resolve has been the evolution and extinction of the Neanderthals. One school holds that Neanderthals, which are now realized to have been restricted to Europe, were ancestral to later, modern Europeans. The other school firmly believes that the change from Neanderthal to modern *Homo sapiens* was so rapid that it must have been caused by a replacement of the Neanderthals by advanced populations moving in from outside Europe, most likely from Africa. Neanderthals in this latter scenario went extinct without issue.

The origin of Neanderthals probably dates from the dispersal of archaic *Homo sapiens* into Europe during the cold and aridity of Isotope Stage 6. One of the main issues that traditionally has confused the situation has been the fact that anatomically modern *Homo sapiens*, already known to have been present in Africa and the Near East, appeared in Europe at the onset of one of the very coldest phases. Why would moderns used to living in warm, Riviera-like conditions on the North African coast want to move into Europe when it was beginning to look like the arctic? The answer to this question has already been given. The Neanderthal replacement was the result of the most recent of a series of population movements of hominids out of Africa that may have begun as early as 2 million years ago. Forced virtually up to the edge of the sea by the spread of desert in North Africa, anatomically modern humans over this vast area found the cold but still well-watered European tundra during the last glacial maximum a more hospitable, and perhaps the only, place to live.

Starting out as animals that clung to the edge of the tropical forest and first ventured forth into the savanna, hominids have come a long way from their roots. Adapting to increasingly more extreme seasonal changes as Pleistocene climate fluctuated, they eventually evolved to tolerate year-round extreme conditions, such as northern Siberia. The cultural attributes of fire, increasingly sophisticated clothing, and increasingly complex tools accounted for this quite rapid ecological transition. Some of these groups, perhaps in the pursuit of game animals or perhaps knowing that they were colonizing new territory, crossed the Bering Strait between Siberia and Alaska at the end of the Pleistocene to populate the Western Hemisphere.

We sometimes forget how close to early hominids we still are. Despite the diaspora of the hominids out of Africa into the glacial cold, we still shiver when our skin is below seventy degrees Fahrenheit. The dietary adaptability that saved our ancestors in periods of near-starvation now allows us to subsist on low-quality junk foods, but at a high cost to our basically early hominid constitutions. Our senses are still attuned to the slightest rustle of a branch or the slightest whiff of an herb, stimuli that our early hominid ancestors found critical to their existence as hunter-gatherers. Yet somehow we tend to forget this ability in the sense-numbing modern world that we have created. As we close in on the missing link of our connection to the natural world in the far past, it is important to begin to assimilate some of the lessons that this search has taught us, for the present and for the future. This is the subject of the last chapter.

# 10

# The Barefoot Species
# with Shoes On

A colleague of mine, Doug Cramer, pointed out that what I had been doing outside the well-worn paths of East African paleoanthropology was countering the "streetlight effect." This effect, he said, could be paraphrased as looking for things not where they might be, but where you could have found them had they been there.

This was exactly what paleoanthropologists had done for years. They looked for fossils only in sites where logistics were easy, where they spoke the language of the country (usually English), where the climate was relatively benign, and where there were good accessible exposures with lots of bone on the surface. Theoretical reasons for investigating certain areas and time periods had never been foremost in designing and planning research expeditions. We have begun to change this approach and thus to ameliorate the influence of the streetlight effect in paleoanthropology.

The streetlight effect received its name from an old Vaudeville act. A policeman comes up to a drunk under a street lamp on a dark night who is obviously looking around on the ground for something he lost. The policeman asks if he can help. The drunk replies, yes, he has lost his glasses. After several minutes of fruitless searching the policeman asks the drunk if he is sure that he lost his glasses here. "No," he replies, "I lost 'em down the street, but the light's mush better here."

Paleoanthropologists are much like the Vaudeville drunk. Through a miasmic cloud, they search for humanity's origin not where it likely occurred but where the deposits are most accessible and the logistics easiest. They then go a step further. If they find something, they try to convince themselves, their colleagues, and the public at large that they have found the missing link. In the Vaudeville metaphor, this would be tantamount to the drunk finding a dollar bill, a woman's purse, or perhaps somebody else's glasses, and trying to convince the policeman that he had found what he had been looking for.

The history of paleoanthropology is replete with examples of this misdirected enthusiasm. Theories of human origin tend to be biased in favor of where one is from. Florentino Ameghino postulated that hominids originated in his native Argentina— an entire hominid lineage going back to the Miocene Epoch, based, as it turned out, on misshapened Indian skulls in incorrectly dated geological strata. Pleasant working environments tend to be favored as postulated sites of human origins. The wine country of southern France, which has had more paleoanthropologists rummaging about and digging holes than any other place on earth, was a favored point of origin until new techniques of radiometric dating made it apparent that all the sites were much too recent to be relevant to early hominid origins. An eminent colleague of mine once told me that he was saving a prime site on a Greek island as an excavation project for his retirement. His wry smile told me that it was more for the cool breezes from the Aegean and the souvlaki than for testing scientific hypotheses.

I have adopted a different philosophy. If there are good scientific reasons for investigating an area, then the more inhospitable it is the more likely it will be to yield something that no one else has ever found. When graduate students or others on my expeditions have complained about harsh conditions, I have told them that if it were easy somebody already would have done it. There was a standing joke among my students that if you wanted to lose weight or be convinced not to go into paleoanthropology, go with me to the field.

There are vast tracts of the scorching and waterless Sahara, hundreds upon hundreds of square miles of potentially fossiliferous sediments, that are virtually unexplored. There are countless unexplored caves in the steamy forests of western and central Africa (not to mention Asia and South America) that may hold entirely unimagined faunas. There are similar paleoanthropological research opportunities in war-torn parts of the world, in parts of the world where investigators do not speak the native languages, where one's own country is not held in the highest esteem, where there are hoards of biting and stinging insects, and where conditions are quite unhealthy. In short, there are hundreds of reasons and excuses for not going to work in any of these places, and they have all been adduced by generations of fieldworkers. This aversion to hardship gives me and my teams quite a bit of opportunity, limited only by resources to send personnel into the field and geopolitical circumstances beyond our control.

Of course, the general applicability of the streetlight effect in paleoanthropology makes me look with jaundiced eye on many of the old hypotheses on hominid origins. I will be ever indebted to Doug Cramer for putting a label on the phenomenon for me. It has been a strong negative factor in paleoanthropology, counteracted only by the serendipity of discovery.

The magnitude of the question of ultimate hominid origins has intrigued us for a long time. The source of humanity as it emerged from the rest of the animal kingdom has eluded discovery ever since Darwin hypothesized its existence almost a century and a half ago. Science has chipped away at the problem, but the central root of the human tree remains undiscovered.

Darwin and Huxley even postulated where this "missing link" should be found: Africa. But for many Westerners, scientists included, Africa has remained a place of foreboding darkness, following the imagery of Henry Morton Stanley and Joseph Conrad. Africa has been the other end of the street, not lit by the street lamp. But Africa is a continent of light—bright sun reflecting off shimmering lakes, long vistas to distant mountains, and the main source of illumination on our ultimate origins.

Africa will one day yield the answer to our origins, a question that is over a century old in its modern scientific formulation. But the quest will be difficult and expensive. It will be an exploratory mission much like the NASA probes to the solar system and beyond. However, when the full-scale probe into human origins takes place, it will discover not the secrets of other life forms but those of our own origins.

## THE MISSING LINK AND OUR VIEW OF THE WORLD

"Big deal," you might say. The missing link is a figure of popular culture anyway. What does it matter if poetic license is exercised in the name of good theater? Of what real importance is knowing the ultimate origins of the human species? To this we might respond that although the question of human origins is as old as philosophy, the answer, when found, will have implications much beyond the confines of anthropology.

Every culture needs to explain its existence. Cultures accomplish this for their members by giving them what sociocultural anthropologists call "existential postulates"—myths and epics that tell people how they were created; how the earth, moon, sun, and stars were formed; and where all the animals and plants came from. Native Americans in the southwestern United States, for example, believed that the Great Spirit had baked people into existence. The Great Spirit had taken some people out too early, and these became white people. Others had been left in the oven too long, and these became black people. But the Great Spirit had baked a third group just the right amount of time. These perfectly complexioned people were the Indians.

Such explanations provide an answer to the ever-questing human mind for the ultimate mystery of human existence. And they have worked for untold millenia. The existential postulates that cultures have used have been critical underpinnings of their overall adaptations. They are the images that people carry when they go about the frequently difficult or even dangerous tasks of making a living.

A culture's existential postulates are the basis for its world view: how it and its members relate to the rest of the world. In the example above, as in all other human cultures, one's own group is viewed as the best. This outlook clearly has been adaptive for cultures and its members. If environmental resources become scarce, the "best" people are clearly entitled to them. If a situation arose in which Native Americans, white people, and black people were competing for the same resources, the Native Americans might choose to deny those resources to the other two groups, and they would have traditional existential postulates to justify their decision. When the reverse actually happened and the Anglo-American culture denied land and environmental resources to the Indians, it was called "manifest destiny."

But the traditional existential postulates and world views of individual cultures no longer work today on a global scale. The world has become a much smaller place, essentially one system with mutually interdependent subunits. What individual societies do affects and concerns everybody else. Consequently, the global community can no longer afford competing, self-serving, and mutually exclusive world views. A single world view must replace the traditional, competitive explanations. People still need an image of themselves and their place in nature to function in their daily lives, but now it must be a consistent, global view.

In the past almost any explanation that satisfied human experience would suffice for a culture's origin myths. The experience and impact of cultures were local and limited. Now there has to be an explanation that transcends the ethnocentrism of individual cultures. We must have a scientifically testable explanation for our origins and our nature. Only then will we be able to construct a world view that adequately meets the needs of all peoples in the global community.

Why is the missing link portion of anthropology's explanation important? After all, many links in the chain of human evolution are already known, and biological anthropology is already making contributions to a wide array of societal concerns.

Like any chain, the chain of reasoning leading to human origins is only as strong as its weakest link. If there is a missing link, then the chain is not complete. We need to know how and when humanity started on its journey out of apedom in order to fill in many of the details of the transition and to paint the overall picture of the evolution of human nature. We know a lot more now than we did just a few years ago, but the human evolutionary chain needs to be anchored. The evolutionary tree of the hominids—our evolutionary tree—needs to be rooted.

## ANTHROPOLOGY, CULTURE, AND HUMAN ORIGINS

Scientists who study people holistically, in their entirety, have been termed anthropologists. Other scientists study specific components of the human organism. Psychologists investigate the workings of the human brain. Anatomists deal with the physical structure of the human body. Physiologists figure out what the structures do. Geneticists tease apart and unscramble the meaning of the DNA-encoded messages that drive human reproduction and development. From an anthropological perspective, these sciences are much like the blind men and the elephant. Each has its own verifiable perspective, be it the tree-like leg, the wall-like side, the snake-like trunk, or the rope-like tail, but only a holistic view will yield the true picture of an elephant.

Long before Darwin, anthropology became obsessed with chronicling the minutiae of human existence in order to attain a comparative, holistic understanding of the human condition. Skulls from remote tribes were measured, languages from all corners of the globe were recorded and transcribed, cultural practices of every society in the world, from kinship terminology to rituals, were written down, and excavations around the world plumbed the depths of the archaeological and fossil records. A mass of data poured in from legions of field anthropologists. Natural history museums filled with collections of the tremendous variety of peoples, a diversity that began to vanish as twentieth-century Western culture spread worldwide and swept all before it.

How did all of this information fit together? How should it be organized? It fell to a German physicist who became fascinated with a holistic view of the human condition to develop the framework. His name was Franz Boas and he became professor of anthropology at Columbia University in New York.

Boas tied all of the data about tribes, languages, skull form, and ancient potsherds together with one paradigm: culture. Culture was defined as that set of shared, learned values and behaviors that characterizes any particular society. In the generic sense culture was viewed as what separates people from animals and what mediates all human behavior.

The cultural paradigm was a powerful concept. It served during much of the twentieth century to organize the academic discipline of anthropology across the United States and in many parts of the world. How else could the study of such widely diverse subjects as sexual mores of South Pacific islanders, the blood proteins of African tribes, the archaeology of Arctic Circle sea-mammal hunters, the linguistic map of the Iberian Peninsula, and the fossil skulls of ancient hominids in China all be subsumed into one scientific field? Culture dictated how members of society acted and learned to act as they grew up, thus determining the sexual mores of the South Sea islanders, and every other culture for that matter; cultural rules determined how members of contiguous tribal groups intermarried, thus explaining the distributions of genetic markers of population relationships; archaeology was no more than the cultural history of the extant peoples of the far north; and languages usually if not always followed the distribution of other patterns of cultural behavior. For example, the Basques were culturally, linguistically, geographically, and even genetically distinct from other inhabitants of the Iberian Peninsula. Culture, language, and genetics (sometimes termed "race" or "biology") seemed to covary. They seemed to be correlated with one another, and this covariance seemed to give anthropology some coherence as a field.

But how did the cultural paradigm incorporate the evidence of extremely ancient fossil relics of humanity? Anthropologists had found that culture provided a critically needed paradigm for pulling together all their field observations on living peoples, and it was natural to project the same structure back into the past.

It was difficult for a number of biological anthropologists and anatomists to accept this cultural paradigm. Their tradition, the true origin of anthropology, lay within the natural sciences. Such an anthropocentric concept as culture ran crosswise to their theoretical disposition and did not particularly help to explain their data and observations. Nevertheless, the general sway of opinion held that ancient hominids, as small-brained and technologically primitive as they appeared to have been, must have been culture-bearing and thus human. The theoretical paradigm only allowed a slower boil within the cultural category, and early hominids, until those discovered in the last decades or so, were clearly more advanced than "animals." Biological anthropologists were swept along with the tide of the cultural paradigm.

## Cultural Origins

Research in two major areas of anthropology have now fundamentally changed the way we have to look at cultural origins. We have already talked about the first of these areas, primatology, the study of the behavior and evolution of nonhuman primates. When Jane Goodall discovered tool use among chimps, the cultural paradigm teetered. The second is archaeology, which has now plumbed the depths of the cultural record back to its earliest beginnings.

Archaeological discoveries and the documentation of cultural evolution extending back over 2 million years further changed our idea of culture. Sociocultural anthropologists during most of the twentieth century had not been interested in the evolution of culture. They had an aversion to studying cultural evolution because of earlier, now discredited, ideas that human cultures all progressed through certain set stages of development from savagery to barbarism to half civilization and finally to civilization. This sequence of evolutionary change was postulated by the mid-nineteenth century social scientist Lewis Henry Morgan, and mention of his name brought a snarl to the lips of all self-respecting ethnologists and sociocultural anthropologists during my graduate school years.

Conventional ethnological wisdom held that individuals and whole cultures could leap from the Stone Age to the Space Age in one or two generations, thus vitiating Morgan's idea of the lock-step evolution of cultural progression. As colonialism brought many of the world's previously isolated cultures into contact with the industrialized West, many examples were known in which individuals and groups in society had rapidly assimilated Western ideas. Morgan's reputation among twentieth-century ethnologists also was not helped by the pejorative implications of the terms he used, such as "barbarism" and "savagery." These designations tended to reinforce the opinion that Morgan was an ethnocentric Euro-American who had uncritically accepted his particular culture as the height of evolutionary advancement. All others were somehow inferior. Contemporary ethnology had adopted the principle of "cultural relativism"— one culture is not better or worse than another, just different.

It now appears that the baby was thrown out with the bath water. Stripping away the language of colonialism and ethnocentrism from Lewis Henry Morgan's thesis reveals that he postulated a series of cultural transformations from simpler to more complex cultural levels through time. His hypothesis suffered from a foreshortened time line because little of great archaeological antiquity was known at the time he wrote. And there was little of the ethnographical wealth of knowledge that was to come in the late nineteenth and twentieth centuries. But Morgan's central thesis of broad stages of cultural progression through time has been amply demonstrated by the archaeological record.

I have been fortunate in being present at the excavation of two of the oldest archaeological sites in the world: Omo,

Ethiopia, investigated by Harry and Joan Merrick, and Senga
5A in the Upper Semliki, Zaire, excavated by Jack Harris. Both
of these sites are well over 2 million years old. The artifacts
that these sites have produced are similar. They are made of
quartz, a notoriously bad type of stone to make tools from
because it is subject to unpredictable breakage caused by the
large cleavage planes that run through it. Also, the tools are all
small flakes. There are very few of the big "cores, " that part of
the rock left when flakes are removed, that were found at
Olduvai Gorge Bed I by Mary Leakey. The hominids at this
very ancient time period, the beginning of stone tool making,
apparently used the very crude method of simply breaking
rocks to obtain sharp shards with which to cut meat and other
objects. They may even have thrown the rocks against other
rocks on the ground to break them up. Calling these crude
pieces of quartz "flakes" may be overly generous because it was
probably beyond the ability of these early hominids to control
the breakage pattern of stones to produce flakes. Lewis Henry
Morgan would surely have called these creatures "beasts," far
below his scale of human barbarism.

The first steps of cultural evolution were so slow and plod-
ding compared to the modern pace of sociocultural change that
it is hard for us moderns to conceive of it as the same process.
It took several hundreds of thousands of years for flaking to be
discovered and used in stone tool making. The difference in
age between Senga 5A (2.3 million years old) and Olduvai Bed
I (1.8 million years old) is about 500,000 years. Superficially the
artifacts appear similar between these two sites, even over this
vast span of time. But there is the subtle difference that controlled single or double flakes were being taken off cores of
rocks at Olduvai. In half a million years hominids had learned
to hold a rock, possibly between their feet, and hit it with
another rock in a controlled way to produce flakes of predictable size and shape. This technique might take a modern
ten-year-old, with the awe-inspiring neural wiring of *Homo
sapiens,* a few hours or a day to figure out. It took *Homo habilis*
500,000 years. There is clearly a qualitative difference in the
culture of early hominids and the culture of modern humans.

The adaptability of culture also was substantially different in early hominids. When populations of *Homo erectus*, the descendants of *Homo habilis*, spread from Africa to the rest of the Old World they encountered very different environments that were colder than any parts of Africa. The flora and fauna they utilized for food and clothing were different. Yet there is no evidence that *Homo erectus* used tools that were substantially different from those they had carried out of Africa. Hand-axes, pointed, pear-shaped tools chipped all around the edges, are essentially identical from southern Africa to southern Asia. They remain unchanged for several hundreds of thousands of years. Again early human culture must have been different and substantially less adaptable to change than our modern cultural behavior.

Culture, then, as a paradigm for understanding human behavior and its evolution became less useful as more and more data came in from archaeology and primatology. Twentieth-century ethnologists coined the termed "superorganic" to describe the nature of modern human culture, which seems to careen off on its own trajectory of change, unfettered by biological evolution. But the earliest phases of culture seem to be much more synchronous with biological change, adapting and changing at a much slower pace. This pace of change began to accelerate only relatively late in the human career.

To understand the nature of culture and to investigate the beginnings of this most important of human social adaptations, anthropologists must adopt a different paradigm, one that provides a framework to test the phenomena in which they are interested. This paradigm is evolution by natural selection. It promises an anthropology that is more evolutionary and less cultural in focus, one beyond relativism, but one that need not look askance at human differences. In short, an evolutionary perspective offers an explanatory link to many of the phenomena that anthropologists seek to explain.

## ANTHROPOGENY: THE STUDY OF HUMAN ORIGINS

Understanding the human condition through recourse to its evolutionary history is a field of research that has begun to be known as "human origins." College and university courses by that name have sprung up across the country in anthropology and biology departments. There is the Institute of Human Origins in Berkeley, California, dedicated to paleoanthropological and geological dating research. And the term has become common in informal usage by anthropologists to connote a multidisciplinary field focused on investigating human biological and behavioral evolution. Perhaps it is time to give a more formal designation to the field.

The great German naturalist Ernst Haeckel was a great coiner of names for new scientific fields of study. He first introduced the terms "ecology" and "phylogeny." Another of Haeckel's terms fell into disuse soon after he proposed it, but now it can provide a descriptive label for the multidisciplinary field of evolutionary anthropology. The term is *anthropogeny*, meaning literally "origin or genesis of humanity." Haeckel proposed it in his 1879 book *Die Abstammung der Menschen* (*The Evolution of Man*).

Anthropogeny has some differences from anthropology, from which it has sprung. The latter has been, since its inception in the eighteenth century, a descriptive field. Anthropologists have described and cataloged physical differences, ancient bones, different cultures, unwritten languages, and exotic artifacts. Now, however, virtually all of the physical, cultural, and linguistic diversity of modern humanity is known. Anthropologists have become more interested in understanding why and how differences and changes occur. They are more analytical, and they focus on specific questions that need to be answered. For example, a researcher might study a tribal group in the Andes to answer questions related to their long-term physiological adaptations to the cold temperatures and low oxygen pressure of high-altitude environments. One such question is whether the barrel-shaped chest of these people is an adaptive response to breathing much larger volumes of air in this environment, an inherited characteristic, or the result of an interaction of both factors. Traditional anthropology would have been content to document that the tribal group had barrel-shaped chests. Anthropogeny is more hypothesis-driven than traditional anthropology, and it is more analytical as a field.

The new field, anthropogeny, uses a paradigm of evolution by natural selection as its framework for investigating problems. The old field, anthropology, uses culture as its paradigm. For example, researchers of early humans in the fossil record do not assume that the subjects of their study had a language, one of the prime components of culture, simply because they were human. This subject requires research. Investigators look at the structure of the brain as evidenced through casts of the insides of ancient skulls and at bony indications of the structure of the vocal apparatus to gain clues about the linguistic abilities of early humans. For traditional anthropology, culture cannot change in its basic adaptive form; for anthropogeny it is a dynamic and evolving entity.

Anthropogeny is a natural science; anthropology is a social science. Because of the difference in their paradigms, these two approaches to understanding human diversity and behavior fall into two different disciplinary categories. Anthropogenic researchers utilize data and techniques that are biological, geological, chemical, and physical, whereas traditional anthropological researchers argue from sociological, economic, and cultural perspectives. Biological anthropology has, of course, spanned these two perspectives, but as more diversity has come into the natural science side of the equation, anthropology has had difficulty assimilating it. For example, very few departments of anthropology have laboratories for the important emerging anthropogenic areas of molecular genetics of human populations or high-technology three-dimensional imaging of human anatomy.

Anthropogeny tends to be more quantitative in approach, using statistics and mathematical tests, whereas traditional anthropology has been decidedly unquantitative. Again only biological anthropology has been an exception, but until the last couple of decades most quantitative studies have involved simple metric dimensions and statistics.

Finally, anthropogeny is much more multidisciplinary than anthropology. Because it is hypothesis-driven, many new fields and techniques can come into play. Anthropogenic researchers may work with protein chemists in developing amino acid racemization dating techniques for investigating emergent human populations 100,000 years ago, or they may collaborate with ecologists to investigate natural savanna ecosystems for comparison with early hominid paleoecology, or they may work closely with computer scientists in developing and utilizing new graphics techniques for investigating and measuring human morphology.

A recent survey of the practicing biological anthropologists in the United States showed that the field is undergoing a transformation toward a more quantitative, more ecologically focused, and more multidisciplinary discipline. The social sciences have dropped precipitously in perceived relevance; sociocultural anthropology and economics are near the bottom of a long list of associated sciences. Anthropogeny in a sense has already arrived on the scene, but it will continue to develop out of anthropology, anatomy, museum, and research institute departments. Anthropogeny, or whatever name it goes by, is a strong, new force in the broad study of humanity, and its evolutionary research agenda will provide profound insights into human nature in the future.

## ANTHROPOGENY AND THE HUMAN CONDITION

What we now know and what we will find out in the near future regarding human origins needs to be utilized. Our species is, indeed, a naked ape dressed up for the ball, a barefoot species with shoes on. Our modern civilization is a very thin veneer when we consider that modern *Homo sapiens* came on the scene 100,000 years ago. The last 500 years, since Columbus's discovery of America, account for a mere 0.5 percent of the last phase of our history as a species. The 1,600 years that have elapsed since the fall of the Roman Empire is only 1.6 percent. This span of sixteen centuries is only about 0.06 percent of the time since the genus *Homo* first appeared 2.5 million years ago. If we go back to a postulated origin of the hominids at, say, 8 million years ago, the veneer of civilization is only at a minuscule 0.002 percent of the human evolutionary experience.

What makes the small proportion of time that we have spent in "civilized" conditions so significant is the magnitude of change in our daily lives that this change effected. Prior to village, semiurban, or urban life, which many of our ancestors may have avoided until only a few decades ago, people lived their entire lives in low population densities with family and a few friends close at hand. Life was physically demanding and hard in the sense that you personally did what got done; little was the product of other peoples' labors. You lived on and from the land, with little excess. There were times of feast and times of famine. Because of the low population density, widespread disease was relatively rare, but when it struck there was no modern medicine, and survival depended on the body's own defenses. Children and infants were hit the hardest and suffered the highest mortality. The people you saw on an average day were those you had grown up with and known their entire lives. Your world was composed of about one-hundred people, living and dead, and your territory was not more than two- or three-hundred square miles. That territory included unpolluted water and air, a lot of open space, quiet, and the surroundings of trees, grass, water, and animals.

What is more than a little amazing is that this basic style of human life, dating back several hundreds of thousands of years at least, changed within the living memory or known family history of most people alive on earth today. The type of life to which we have been evolutionarily adapted has changed overnight, and we now find ourselves in a very foreign environment, although one of our own making. Can there be any doubt that there will be significant and severe repercussions to the quality of life and that individual human beings will struggle desperately to regain a sense of adaptation to a personal world gone awry?

Modern civilization is a strong force. It pulls us along with it whether we like it or not. The old cultural anthropological description of culture as being superorganic has taken on a more sinister meaning. It can now connote the treading underfoot of human life by a culture run amok, instead of indicating, as it was originally intended, cultural change occurring over, or more rapidly than, biological change. Author Jules Henry's phrase "culture against man" encapsulates this idea.

Applied anthropogeny can help us re-establish some balance

between our biological selves and our hyperactive sociocultural milieu. We need to ensure a fit between our organic and super-organic identities. Both are products of the evolutionary process. By studying and understanding that process, we will be able to interface much more effectively our frequently opposing biological and psychosocial selves. No other area of applied anthropogeny has greater importance than that of medicine and health.

## EVOLUTIONARY MEDICINE

Human health has been one of the first areas in which the idea that culture acts against our best long-term interests has taken hold. There is a movement afoot called "evolutionary medicine" that takes as its working paradigm the idea that human beings are the product of a long evolutionary process. Conditions of normal modern life, and medical treatments designed to return us to those conditions, must reflect a fundamental awareness of our adaptations as a species.

This way of looking at human disease and dysfunction seems logical to many, but it has been resisted by others. Those who resist the message of evolutionary medicine have a different paradigm, based on mechanical and engineering principles. To them the human body is a machine for which history is irrelevant. Does it matter whether a car engine is a Ford or a BMW, made in Detroit or Munich? They all work according to the principles of the internal combustion engine. Thus if the fuel pump fails, you replace the components, the valves, or the entire fuel pump. It's as simple as that.

Transferred to humans, this philosophy works in a similar way. Patient Jason Smith has been diagnosed with severe heart disease and admitted to the hospital. Translation: his fuel pump is worn out. Solution: Replace the components, put in some new valves, or replace the entire pump—transplant the heart. It's as simple as that.

But it's not as simple as that. Artificial heart valves, coronary bypasses, pacemakers, and transplanted hearts do not work as well as the originals. The patients usually have a substantially reduced quality of life. Clearly something is wrong with the paradigm. But a doctor's options are limited when a patient comes in with an advanced disease. This example shows up how failed the engineering paradigm in medicine is. If one waits for mechanical failure, it is too late.

The approach of evolutionary medicine is less invasive, more preventive, less all-powerful for the doctor, and more empowering for the patient. It is also much less expensive, and for that reason alone policy makers and insurance companies should be very interested in it.

Evolutionary medicine came about through the realization that there are "diseases of civilization," an expression coined by Boyd Eaton and Melvin Konner, both M.D.-anthropologists at Emory University in Atlanta. Diseases of civilization include the scourges of modern life: cancer, heart disease, high blood pressure, and gastric ulcer. These maladies are virtually unknown in traditional hunter-gatherer societies. What in modern life causes these diseases? Can we learn from human evolution how to deal more effectively with them?

The answer is an unequivocal yes, even though obviously we cannot return to living *au naturel* on the African savanna. The first step is to chronicle the differences in lifestyle between our ancestors, the conditions to which we are ideally adapted, and the conditions in which we find ourselves today.

We know from the fossil plants found associated in the fossil record with early hominids that the environments of our ancestors were seasonal, with a clear rainy season and a clear dry season. In equatorial Africa, when there is rain there is food, and when there is drought food is scarce. Hominids adapted to this seasonal round by developing the physiological mechanism of fat storage. Hominids store up energy in the form of fat during the season of plenty and utilize this fat during the season of low food availability.

But consider the modern Western environment. There are no dry seasons at Safeway. We live in an environment of perpetual plenty. People eat foods supercharged with calories that are so refined and so easy to swallow that we hardly know that we have chewed anything and eaten it. Early hominids ate a tremendous amount of unrefined fiber in their food, and our digestive systems are still adapted to this diet. Our tastes for sugar, salt, and fat, substances that were limited by the environment during our evolutionary infancy, can now be indulged to the fullest with candy bars, potato chips, and pizza. Our fat stores are never used because the lean season that our bodies expect never arrives. Obesity has become a major problem. Arteries become clogged with plaque caused by high cholesterol in our blood, in turn contributing to high blood pressure, stroke, and heart attack. Abnormally high sugar levels contribute to diabetes. The lack of dietary fiber causes intestinal diverticulitis and has been implicated in various intestinal cancers. And general overweight conditions cause such diverse problems as flat feet and bad backs because our skeletal frames are not designed to support the disproportionate weight that obese Westerners carry.

Research on the diet of early hominids, perhaps through analysis of their fossilized feces, probably has more to contribute to preventive medicine than many of the studies currently undertaken at great cost. As yet this avenue of research is only beginning to open up. Further work on the paleoecology of early hominids can also contribute significantly to an understanding of the contexts of our own physiological adaptations.

## HUMAN ECOLOGY AND ADAPTATION

Early hominids evolved under conditions substantially different from those in which they frequently find themselves today, a situation that not surprisingly causes problems in adaptation. Yet the realization that human evolutionary principles can materially aid in the interpretation of the problems and contribute to solutions still eludes many of us.

Perhaps one of the classic cases in which a natural science perspective on human behavioral problems became lost in a flurry of misunderstanding involved Dr. Fred Goodwin, former director of the Alcohol, Drug Abuse, and Mental Health Administration in Washington. In April 1992 Goodwin suggested in a meeting that field research on rhesus monkeys might reveal important insights into an area of human behavior that is of concern to us all, urban violence and crime. Noting that young rhesus monkeys become hyperaggressive when they leave their troop at adolescence, he suggested a comparison with inner-city young males who have lost the social fabric of family and school. But Goodwin was perhaps too informal when he noted that it "is the natural way of it for males, to knock each other off" and "maybe it isn't just the careless use of the word when people call certain areas of certain cities 'jungles'". The Congressional Black Caucus called for Goodwin's immediate dismissal on the basis that he had made racist comments. Senator Edward Kennedy and Representative John Dingell called using nonhuman primates "a preposterous basis" for studying human behavior. Bowing to the pressure, Goodwin resigned.

This unfortunate incident underscores the immense gulf of ignorance that still yawns before us. Until people can think of themselves as members of one species, as relatives all descended from a common source, as part and parcel of nature, in ethological balance with their biological heritage, and in ecological balance with the rest of the world's resources, we will have little chance of solving many of society's ills.

This need to reach this understanding is why the missing link is important. When it is found and the circle is closed, there will no longer be a reason to set humanity apart from nature. An unbroken chain will stretch from "animal" to "human," and we will realize that there is much shared between the two. To deny our specific animal origins and nature is paradoxically to deny the uniqueness of our humanity. We have much to learn.

# ACKNOWLEDGMENTS

My debt to the colleagues whose work has helped to form and direct my own research is immeasurable. Sherwood Washburn, Charles Kaut, Clark Howell, Jean de Heinzelin, Frank Brown, Alison Brooks, Jack Harris, Ali El-Arnauti, Dorothy Dechant Boaz, Hank Wesselman, Raymonde Bonnefille, Kay Behrensmeyer, and the late Glynn Isaac, Wahid Gaziry, and Doug Cramer, taught me a great deal and have all contributed in their own important ways to the development of the science and the ideas presented in this book. I have also learned much from those with whom I have disagreed, Tim White, Don Johanson, and Richard Leakey. For institutional support I owe much to the International Institute for Human Evolutionary Research and George Washington University, particularly Tony Coates, Irwin Price, and Ed Jones, without whose interest and support much of the work reported here could not have been completed. I thank my editor Bruce Nichols for many good discussions and his invariably good suggestions. Finally, this book would never have seen the light of day if not for my wife Meleisa McDonell, whose encouragement inspired the whole project and whose editorial acumen got me through quite a few impasses.

# INDEX

*Africa*

Sahabi ☆
LIBYA

*Niger R.*

*Nile R.*

Hadar ☆
ETHIOPIA

Omo ☆
East
Turkana
*Zaire R.*

Semliki ☆ UGANDA
KENYA

ZAIRE
Olduvai ☆
Laetoli
TANZANIA

Sterkfontein
☆

Taung
☆

SOUTH
AFRICA

0          1000 km